Qian Wang
Editor

Chinese Philosophy of Technology

Classical Readings and Contemporary Work

Springer

Editor
Qian Wang
Department of Philosophy
Dalian University of Technology
Dalian, Liaoning, China

ISSN 1879-7202 ISSN 1879-7210 (electronic)
Philosophy of Engineering and Technology
ISBN 978-981-15-1951-2 ISBN 978-981-15-1952-9 (eBook)
https://doi.org/10.1007/978-981-15-1952-9

This Springer imprint is published by the registered company Springer Nature Singapore Pte Ltd.
The registered company address is: 152 Beach Road, #21-01/04 Gateway East, Singapore 189721, Singapore

Philosophy of Engineering and Technology

Volume 34

The *Philosophy of Engineering and Technology* book series provides the multifaceted and rapidly growing discipline of philosophy of technology with a central overarching and integrative platform.

Specifically it publishes edited volumes and monographs in:

- the phenomenology, anthropology and socio-politics of technology and engineering
- the emergent fields of the ontology and epistemology of artifacts, design, knowledge bases, and instrumentation
- engineering ethics and the ethics of specific technologies ranging from nuclear technologies to the converging nano-, bio-, information and cognitive technologies
- written from philosophical and practitioners' perspectives and authored by philosophers and practitioners

The series also welcomes proposals that bring these fields together or advance philosophy of engineering and technology in other integrative ways.

Proposals should include:

- A short synopsis of the work or the introduction chapter
- The proposed Table of Contents
- The CV of the lead author(s)
- If available: one sample chapter

We aim to make a first decision within 1 month of submission. In case of a positive first decision the work will be provisionally contracted: the final decision about publication will depend upon the result of the anonymous peer review of the complete manuscript. We aim to have the complete work peer-reviewed within 3 months of submission.

The series discourages the submission of manuscripts that contain reprints of previous published material and/or manuscripts that are below 150 pages/75,000 words.

For inquiries and submission of proposals authors can contact the editor-in-chief Pieter Vermaas via: p.e.vermaas@tudelft.nl, or contact one of the associate editors.

More information about this series at http://www.springer.com/series/8657

Contents

Chapter 1
Introduction

Qian Wang

Research in the philosophy of technology is manifested in diverse theoretical frameworks associated with different technical practices and related to different social contexts. The philosophy of technology as a systematic academic discipline emerged in China in the 1980s. Changshu Chen (陈昌曙) (1932–2011), a prominent philosopher and professor at Northeastern University in Shenyang, took the lead in applying philosophical research to technical activities, and was dedicated to exploring philosophical concerns reflected in modern Chinese technological development. Following his model, many Chinese scholars have been introduced to this field of research and have made constructive academic contributions. The rise of Chinese philosophy of technology has a special historical background and significance. This is worth the consideration of the international academic world.

According to the work by Joseph Needham, ancient Chinese engineering technology was highly developed from the 3rd to 13th century (Needham, 1954, pp. 3–4). But it began to fall into decay at the end of the Ming Dynasty and the beginning of the Qing Dynasty. Modern science and technology in China did not emerge domestically, but came from the West. Thus, modern Chinese science and technology has developed in a social environment where there is a constant blending and clashing of Chinese and Western culture. Chinese philosophy of technology issues differ from those of other countries and regions. Traditional Chinese philosophical ideas and concepts are influenced by Confucianism, Daoism, Legalism and the School of Guan Zi (管子), with "humanistic-pragmatism" value orientation and an emphasis on "conforming to nature". Chinese technicians focused on the creative function of intuition and experience, exploring people's physiological and physical potential, which allowed ancient Chinese technology based on agriculture and handicrafts to reach the top level in the world with a large

Q. Wang (✉)
Department of Philosophy, Dalian University of Technology, No. 2, Linggong Road, Ganjingzi District, Dalian, China
e-mail: qianwang@dlut.edu.cn

Q. Wang (ed.), *Chinese Philosophy of Technology*, Philosophy of Engineering and Technology 34, https://doi.org/10.1007/978-981-15-1952-9_1

number of influential inventions. However, this over-reliance on the role of intuition and experience meant that logical thinking and scientific experiments did not play a substantial role in the development of ancient China's technology. This is the main reason why the system of traditional technological theories failed to become a system of modern technological theories, and also an important factor that causes "Needham's Puzzle". Some scholars have interpreted the social impact and cultural implication of the modern technology that came from the West. In the 1920s, an influential "debate between science and metaphysics" broke out in Chinese academia concerning the issue of social influence of modern science and technology. Chinese philosophers such as Youlan Feng (冯友兰) (Feng, 1985), Shuming Liang (梁漱溟) (Liang, 1989), Zehou Li (李泽厚) (Li, 1986), and other scientists including Sicheng Liang (梁思成) (Liang, 2016) and Xuesen Qian (钱学森) (Qian, 1957), have mentioned the social influence of modern technology and technological aesthetics in their works. Their discussions and ideas can however not yet been viewed as professional research into the philosophy of technology.

Research that comes closer to philosophy of technology are the studies of the Dialectics of Nature that began in the 1950s and concerned the analysis of the development law of science and technology from the perspective of Marxist Philosophy and its ideological influence. In the 1980s, the academic features of these studies of the Dialectics of Nature were strengthened and the ideological features weakened along with the reform and open policy that had been implemented in China. The philosophical issues specific to the technological development in modern China were explored in an academic exchange between China and the West, with deeper understanding of the research into the European and American philosophy of science and philosophy of technology. This exchange created a necessary condition for the emerging of modern Chinese philosophy of technology as a discipline. A series of important readings were published, which became the landmarks of modern Chinese philosophy of technology. Examples are the article "A Brief Discussion of Research into Philosophy of Technology" (Chen, 1983), the books *On Technology* (Yuan & Chen, 1986), *Science and Technology Studies* (Chen, 1985), concerned the essential characteristics of technology, the social influence of technology as well as the comparison between Chinese technological development modes and that of the West. In this period also the Professional Committee of Philosophy of Technology was established within the Chinese Society for Dialectics of Nature. The professionals in the field of Chinese philosophy of technology mainly come from universities of science and engineering, and many of them have the possibility to communicate and cooperate with technological experts and enterprise administrators and create collaborative work on the philosophical aspects of practical issues in the development of modern engineering and technology.

In the 1990s, academic interactions within this field were further developed between Chinese and Western scholars. Chinese scholars have translated and commented on some important classic works written by Western scholars, such as Carl Mitcham (Mitcham, 1999), Martin Heidegger (Seubold, 1993), Lewis Mumford, Herbert Marcuse, and Jürgen Habermas (Gao, 1996), conducting comparative studies in this field. Meanwhile, Chinese scholars began to develop interest in specific problems concerning the philosophy of technology in Chinese cultures, including evolution of Chinese technological ideas (Wang, 1995), the influence of the Chinese cultural environment on importation of advanced technology and technological inventions (Yuan, Chen, & Wang, 1994), the cultural condition of Chinese technological modernization (Chen, 1995).

Since the turn of the 21st century, Chinese scholars have developed their own academic focus and approach, yielding remarkable accomplishments. Their work in the field of philosophy of technology is characterized by the following perspectives:

Firstly, the research into the philosophy of technology is conducted with the focus on technological functions and values, exploring the evolution laws of technology and discussing the interactive relationship between technological development and social factors. The need of Chinese technological modernization emphasized the necessity of technological functions and values, while traditional Chinese culture was used to stress the application of technology ("pragmatic reason" named by Zehou Li), as well as the methodology of the organic relations of things. These factors made Chinese scholars forming some unique ideas. Professor Changshu Chen is thought to be the representative scholar in this field with his accomplishment *Introduction to Philosophy of Technology* (Chen, 1999). Influenced by Professor Changshu Chen, some scholars have conducted further research into technological functions and value properties, such as the investigation into the modern value of the relationship between "*Dao*" (道) and "*Ji*" (技, technique) (Wang, 2009), the discussion of the technological socialization process (Chen & Zhang, 2002), the work on the relationship between technological reason and human values (Zhao, 2001), and the exploration of the aesthetic connotation of engineering technology (Zhu & Shao, 2017).

Secondly, attention is paid to the philosophical issues of technological evolution and innovation with the consideration of the characteristics of China's technological modernization. The representative scholars are Professor Deyu Yuan (远德玉) from the Northeastern University and Professor Shixu Guan (关士续) from Harbin Institute of Technology. Professor Deyu Yuan proposes the way of looking into the laws of technological evolution from the perspective of the process theory (Yuan, Ding, & Ma, 2005). And Professor Shixu Guan emphasizes the philosophical research into technological innovation based on the need of enterprise (Guan & Wang, 2001). Their research perspectives have been employed in a series of relevant studies.

Thirdly, the research into the relationship between ethics of technology and professional ethics are stressed. Since the beginning of the 21st century, Chinese scholars have conducted the research into ethics of technology systematically. In 2003, "Sino-German Symposium on the Ethics of Science and Technology", advocated by Professor Zeyuan Liu (刘则渊) from Dalian University of Technology and Professor Hans Poser from Technical University of Berlin, was held. This marks when Chinese scholars' research into ethics of technology engaged with the international communication. From then on, many Chinese scholars have carried out studies of ethics of technology in reference to the practical issues of China's technological development and traditional Chinese ethical theories, such as the research into ethics of technology (Zhao, 2005), ethics of engineering (Xiao, 2009), the history of ethics of technology, and the education of ethics of science and technology (Cao, 2004). It is worth mentioning that important progress have been made in probing hi-tech ethics (such as Nano-ethics), life ethics, internet ethics as well as the ethics of cognition science (Han & Ouyang, 2010).

Fourthly, the exploration of engineering philosophy has been given great concern. Professor Bocong Li (李伯聪) from the Chinese Academy of Sciences University proposes the "triad theory" that views "philosophy of engineering" as an independent subject to be studied. His representative work is *Introduction to Philosophy of Engineering* (Li, 2002). This is a distinctive research direction of China's philosophy of technology. Being guided by his work, some scholars have carried out research into philosophy of engineering, engineering methodology (Yin, Wang, & Li, 2007), and China's contemporary history of engineering (Li, Li, & Li, 2017), which has resulted in some remarkable achievements.

Except for the four distinctive directions mentioned above, many other scholars have also reached important achievements in the academia of philosophy of technology in China. Some Chinese scholars have produced notable works reviewing and researching Western philosophers and ideological schools of philosophy of technology, such as the research of Ernst Kapp, Friedrich Dessauer, Heidegger, Andrew Feenberg, Don Ihde, Bernard Stiegler, and Jacques Ellul. Others have employed the paradigms of Western schools, such as analytical philosophy of technology, phenomenological philosophy of technology, and pragmatist philosophy of technology, to look into some practical issues. There have been some achievements in this area. Moreover, some scholars have investigated the general philosophical issues in emerging technological fields, such as information technology, artificial intelligence, biotechnology and nuclear technology. Specifically, Professor Huaxia Zhang (张华夏) and Professor Zhilin Zhang (张志林) from Sun Yat-sen University are concerned about technological cognition, technological inference, technological explanation (Zhang & Zhang, 2005). Professor Guosheng Wu (吴国盛) from Peking University focuses on the phenomenological philosophy of technology (Wu, 2009). These scholars have some achievements in their fields. Additionally, Professor Feng Xiao (肖峰) from South China University of Technology has worked in the field of the philosophy of information technology and proposes some important ideas (Feng, 2017). It is worth mentioning that in the field of modern Chinese philosophy of technology, some emerging young scholars have also carried out distinctive studies and arrived at valuable results in the philosophy of cognitive technology, ethics of artificial intelligence, life ethics, technical risk, practical effectiveness of engineering ethics, and responsible innovation. For example, Professor Weiwen Duan (段伟文) from the Chinese Academy of Social Sciences looks into technological rationality (Duan, 2001), Professor Lun Li (李伦) from Hunan Normal University investigates into internet ethics (Li, 2007), and Professor Qingfeng Yang (杨庆峰) from Shanghai University is interested in Don Ihde's phenomenology of technology (Yang, 2015). These studies are thought to have great research potential.

Generally, Chinese scholars have expanded these research dimensions and enriched theoretical work in this field since the 1980s, making indispensable and valuable contributions to a diversified development in the philosophy of technology. This has been briefly introduced in some English journals such as the review articles from Yuanlin Guo (郭元林) (Guo, 2014), Nan Wang (王楠) (Wang, 2015) and Guihong Cao (曹桂红) (Cao forthcoming). However, the accomplishments of Chinese philosophy of technology have not yet been fully recognized within international academia due to the language barrier since these works are rarely written in English or translated into English for publication. Hopefully, this barrier will be bridged for further development both in China and international academic exchanges in this field. With the support and encouragement of Professor Pieter E. Vermaas, who is the editor-in-chief of the series, *Philosophy of Engineering and Technology*, and Professor Carl Mitcham, we compiled this volume in order to further generate international understanding of Chinese research and to improve academic exchanges in the philosophy of technology.

This volume consists of classic readings and contemporary work in Chinese philosophy of technology. The classic readings were written by the first generation of scholars in this field. They were dedicated to leading the direction of modern Chinese philosophy of technology. In the process of selecting the classical readings for this volume, attention is drawn to the achievements that continue to play a significant role in modern Chinese philosophy of technology today. Also, the authors of the readings (except Professor Changshu Chen who has sadly passed away) were able to supplement and modify their articles. This is also the reason why works by some scholars of philosophy of technology from the first generation have not been included. The contemporary work included in the volume is written by second generation Chinese philosophers of technology. Their work reflects the characteristics of research achievements by contemporary Chinese scholars. The "contemporary work" part does not include the work of Chinese scholars who employ Western research paradigms, methods, and reviewing and researching Western philosophers and ideological schools of philosophy of technology that Western scholars are acquainted with, since this volume aims to inform the academic work by Chinese scholars in the philosophy of technology that is less accessible to the English speaking community. The "contemporary work" does not choose the young scholars' work because that their academic career needs to be further developed, and their research results need to be evaluated systematically in the future. Both classic readings and contemporary work collected in this volume are representative to the original and creative research achievements of Chinese scholars in the field of Chinese philosophy of technology. By emphasizing theoretical and cultural features in Chinese academia, this volume might not be as comprehensive as it was intended to be, but nevertheless, it is an attempt to present an overall landscape of academic development in the field of Chinese philosophy of technology.

The "classical readings" part in this volume contains five articles from the first generation of scholars in Chinese philosophy of technology. They are presented corresponding to the four research perspectives of China's contemporary philosophy of technology mentioned above.

The first article, "A Brief Discussion of the Research into Philosophy of Technology" published in 1983 comes from Professor Changshu Chen. It is the first article on modern Chinese philosophy of technology with great value. Not only does it put forward the basic issues that modern Chinese philosophy of technology needs to be concerned with, but also indicates its research approaches and methods. Its original features are kept because of the consideration of its historical value and innovative significance.

The second article comes from my paper "The Features and Values of Changshu Chen's Philosophy of Technology Ideas" for the 19th SPT conference held in 2015. Because Professor Chen passed away in 2011 and could not summarize his theoretical ideas himself, this article analyzes and comments on the theoretical features and academic value of Chen's ideas of philosophy of technology. Based on Professor Chen's main writing, this article points out that the theoretical feature of Chen's ideal is attaching great importance to aspects ranging from analyzing technological activities in terms of function and value, to analyzing the dynamics and interactions among related components in technological activities, to assuming a problem-oriented approach. The academic value of Chen's work is that it offers research paradigms and approaches that are applicable to the technological development of China. Moreover, his work underlines the rules of development for technological activities through comparative studies between Chinese and Western research in this field. Chen's work helps to promote the thinking ability of Chinese technicians, and helps to inform international academia of Chinese research characteristics within philosophy of technology, leading to better communication worldwide.

The third article is "Technology and Its Values from the Perspective of Process Theory", written by Professor Deyu Yuan. The perspective of process theory in this article does not come from Whitehead's process philosophy, but is related to the concept of motion and change in materialistic dialectics as well as to the idea of "continuous reproduction in endless succession" from Chinese traditional culture. Technological activities are viewed as a process and change in a continuous way, on the basis of which the different forms and values of technology are analyzed.

The fourth article "Ethics of Technology and Engineer's Professional Ethics" is co-authored by Professor Zeyuan Liu and Professor Guoyu Wang (王国豫). Professor Liu's outstanding contribution is considered as initiating research on ethics of technology in China. This article is one of the earliest works in the ethics of technology and has exerted great influence on research and education in ethics of technology in China.

The fifth article "The Triad of Science, Technology and Engineering", was written by Professor Bocong Li. As an innovator in the study of philosophy of engineering in China, Professor Li introduces in this article his "triad" paradigm of "science-technology-engineering" as well as his creative ideas about values and ethics of technology.

For the "contemporary work" part in this volume, eight articles are selected as representative of contemporary scholars, and they present continuous achievements of the research perspectives of China's contemporary philosophy of technology mentioned above. What these eight articles have in common is that they are directed towards practical issues of universal significance in traditional Chinese culture and the development of modern society, employing the ideas and methods of philosophy of technology, and proposing new research perspectives, theoretical frameworks and conclusions. These articles are enriching the theoretical system and practical applications of philosophy of technology. What follows is an introduction to these eight articles.

The first article is "Aesthetic Factors in Engineering Activity", written by Professor Baowei Zhu (朱葆伟) from the Chinese Academy of Social Sciences. This article is representative of Chinese scholars' unique comprehension of engineering aesthetics within Chinese culture. The second article is "The Sinicized Process of Philosophy of Technology and the Research in Socialization of Technology" written by Professor Fan Chen (陈凡) and his student, Associate Professor Jia Chen (陈佳), from Northeastern University. A great deal of historical data about the evolutionary relationship between the studies in Dialectics of Nature and modern Chinese philosophy of technology are introduced in this article. Professor Fan Chen has made many contributions to promoting localization of the studies in philosophy of technology in China with some original opinions in socialization studies of technology. The third article is "Thinking about the *Dao* of Technology: A Philosophy of Technology from the Perspective of Chinese Culture," written by myself and my student, Associate Professor Wei Zhang (张卫) from Central China Normal University. This article interprets the essential characteristics of technical activities based on the relationship between the traditional Chinese concepts of "*Dao*" and "*Ji*". By comparing these concepts to the Western philosophy of technology, this article also illustrates the value and contemporary significance of the traditional Chinese concepts of "*Dao*" and "*Ji*". The fourth article is "The Developing Process of Technological Rationality and Its Humanistic Relation", written by Professor Jianjun Zhao (赵建军) from the Central Committee Party School of the Communist Party of China. In the article, he divides the development of technological rationality into three stages, i.e., implement rationality, operation rationality and value rationality, discussing technological pessimism as a representative viewpoint in technological rationality critique from a humanistic perspective. Where the issues discussed in the first four articles are more general, following the research direction of the functions and values of technology concerned by China's contemporary philosophy of technology, the final four articles focus on relatively specific issues.

The fifth article is "Three Dimensions of Nano-ethics," written by Professor Sanhu Li (李三虎) from Guangzhou Institute of Public Administration. This article analyzes theoretical and practical issues in the study of nano-ethics in three dimensions, that is, practical exploration, human experience, and ethical governance. It also represents the contributions of Chinese scholars to the study of higher technology ethics. The sixth article is "Technology, Innovation and Life world: A Phenomenological Analysis", written by Professor Dazhou Wang (王大洲). This article mentions some distinctive works regarding the philosophy of technological innovation by Chinese scholars. Professor Dazhou Wang was the student of Professor Shixu Guan. Dazhou Wang' research in philosophy of technological innovation is guided by Professor Shixu Guan with Chinese scholars' specific research direction and achievements. The seventh article is "The Philosophy of Information Technology and Its Contemporary Value", written by Professor Feng Xiao (肖峰) from South China University of Technology. This article conducts a distinctive analysis of the essential characteristics of the philosophy of information technology, and demonstrates its impact on contemporary social development from the perspective of Chinese scholars. The eighth article is "Approaches to Nano-ethics and the Chinese Perspective", written by Professor Guoyu Wang and her student, Dr. Lei Li (李磊). By conducting a comparative study of Chinese and Western ethics of technology, this article analyzes ethical problems of nanotechnology in China and its social impact on the development of nanotechnology. These last four articles focus on the research directions of technological ethics, philosophy of technological innovation and philosophy of information technology accordingly in China's contemporary philosophy of technology.

Research in the philosophy of technology in China is still in progress, yet is developing to a degree of maturity with global influence. The driving forces behind China's contemporary development in the philosophy of technology come from rapid economic and technological advancement, from the fecundity of traditional Chinese culture (i.e., Daoism, Confucianism, Legalism, the school of Guan Zi as well as the philosophy of neo-confucianism in the Song Dynasty and the Ming Dynasty), and also from the inspiration of ideological exchanges with various international schools in this field. In the contemporary study of the Chinese philosophy of technology, some scholars pay more attention to comprehensive, intuitive, and experiential perspectives in technological cognitive activities with the focus on interpreting the social function of technology based on axiology and process theory. These perspectives are important supplements for Western schools in this field, as Western research stresses analytical, logical, and inferential perspectives which conform to epistemology and methodology. The study of Chinese philosophy of technology still needs improvement and more systematic research methods, and is expected to develop into a more integrated and detailed theoretical system. Future generations of Chinese scholars will strive to make greater contributions to international research of the philosophy of technology.

References

Cao, N. (2004). Reflection on education of engineering ethics in chinese colleges and university. *Research in Higher Education of Engineering, 5*(37–39), 48.

Cao, G. (forthcoming). The reception of western philosophy of technology in China. *Techné: Research in Philosophy and Technology.*

Chen, C. (1983). A brief discussion of research into philosophy of technology. *Journal of Northeast Institute of Technology*, 1.

Chen, C. (1985). *Science and technology studies*. Hangzhou: Zhejiang Education Publishing House.

Chen, F. (1995). *Introduction to technology socialization*. Beijing: China Renmin University Press.

Chen, C. (1999). *Introduction to philosophy of technology*. Beijing: Science Press.
Chen, F., & Zhang, M. (2002). *Analysis of technology*. Fuzhou: Fujian People's Publishing House.
Duan, W. (2001). *Truss up time: Technology and human life world*. Guangzhou: Guangdong Education Publishing House.
Feng, Y. (1985). *The brief history of Chinese philosophy*. Beijing: Peking University Press.
Gao, L. (1996). *Technology in the perspective of humanities*. Beijing: Chinese Social Science Press.
Guan, S., & Wang, D. (2001). Several questions about technology innovation. *Journal of Northeast University (Social Science Edition), 2,* 115–118.
Guo, Y. (2014). The philosophy of science and technology in China: Political and ideological influences. *Science & Education*, 23.
Han, D., & Ouyang, k. (2010). *Puzzles and frontier: Studies on ethical problems of science and technology*. Beijing: People's Publishing House.
Li, Z. (1986). *On the ancient chinese thought of history*. Beijing: People's Publishing House.
Li, B. (2002). *Introduction to engineering philosophy*. Zhengzhou: Daxiang Publishing House.
Li, L. (2007). *Ethics of network communication*. Changsha: Hunan Normal University Press.
Li, B., Li, S., & Li, B. (2017). *An outline history of modern engineering in china*. Hangzhou: Zhejiang Education Publishing House.
Liang, S. (1989). *The complete works of Liang Shuming*. Jinan: Shandong People's Publishing House.
Liang, S. (2016). *Chinese building and art*. Beijing: Beijing Publishing House.
Mitcham, C. (1999). *An introduction to philosophy of technology* (Y. Dengxiang, C. Nanyan, et al., Trans.). Tianjin: Tianjin Science and Technology Press.
Needham, J. (1954). *Science and civilisation in china. Introductory orientations*. Vol. I, Cambridge: The Syndics of the Cambridge University Press.
Qian, X. (1957). Methodological problems in technical science. *Journal of Research and Communication of Dialectics of Nature*, 1–34.
Seubold, G. (1993). *Heideggers analyse der neuzeitilichen Technik* (S. Zhuliang, Trans.). Beijing: Chinese Social Science Press.
Wang, Q. (1995). *Board and profound scientific thought*. Shenyang: Liaoning Ancient Books Press.
Wang, Q. (2009). *Between Dao and technique: The philosophy of technology in chinese cultural background*. Beijing: People's Publishing House.
Wang, N. (2015). Introduction for SI:EPEED on TIS the emergence of Chinese discussions of engineering. *Philosophy and Ethics, Technology in Society, 43,* 52–56.
Wu, G. (2009). *Lectures on philosophy of technology*. Beijing: China Renmin University Press.
Xiao, F. (2017). *Information technology in the view of philosophy*. Beijing: Science Press.
Xiao, P. (2009). *Introduction to ethics of engineering*. Beijing: Peking University Press.
Yang, Q. (2015). *Albatross soaring: studies on Don Ihde's technological phenomenology*. Beijing: China Social Science Press.
Yin, R., Wang, Y., & Li, B. (2007). *Philosophy of engineering*. Beijing: Higher Education Press.
Yuan, D., & Chen, C. (1986). *On technology*. Shenyang: Liaoning Science and Technology Press.
Yuan, D., Chen, C., & Wang, H. (1994). *Comparison of technological innovation between China and Japan enterprises*. Shenyang: Northeastern University Press.
Yuan, D., Ding, Y., & Ma, Q. (2005). *Industrial technology studies*. Shenyang: Northeastern University Press.
Zhang, H., & Zhang, Z. (2005). *Studies on technological interpretation*. Beijing: Science Press.
Zhao, J. (2001). *Inquiries into technological pessimism*. Shenyang: Northeast University Press.
Zhao, Y. (2005). *High-tech ethics*. Shenyang: Northeast University Press.
Zhu, B., & Shao, Y. (2017). Aesthetic factors in engineering activities. *Studies in Dialectics of Nature, 3,* 57–62.

Part I
Classical Readings

Chapter 2
A Brief Discussion of Research in Philosophy of Technology

Changshu Chen

Abstract The philosophy of technology should be the philosophy which guides human's remaking nature, as well as the directing principle for the achievement of modernization of science and technology. We should research and establish a philosophy of technology that exhibits scientific and Chinese characteristics while adopting theories from Western philosophy of technology and Japanese technology studies for reference. The philosophy of technology should focus on the practical problems concerned the relationship between technology and nature, between technology and society, the relationship among the internal element of technological development as well as the methodology of technological invention and application, in particular the problems concerned the value and evaluation of technology.

Keywords Philosophy of technology · Oriented problem · Value · Evaluation

2.1 Establishing the Philosophy of Technology with Chinese Characteristics

The term "philosophy of technology" is by no means a novel phenomenon. As early as 1877, Ernst Kapp, a German scholar named, wrote *Grundlinien einer Philosophie der Technik*. In 1914, Eberhard Zschimmer, another German scholar, published a

---

Changshu Chen is deceased.

Explanatory note: this paper was published by Professor Changshu Chen in 1983 and it is the first one as classic reading in modern research of philosophy of technology in China. Some of its references are in Chinese translations of foreign literature and these translations had failed to keep all of the information of their original works at that time. Since Professor Chen has passed away for years, it's impossible for us to get the information about the original works of these Chinese translations. Hence, we translate Professor Chen's Chinese manuscript into English in accordance with the original.

C. Chen (✉)
Northestern University, Shenyang, China
e-mail: qianwang@dlut.edu.cn

Q. Wang (ed.), *Chinese Philosophy of Technology*, Philosophy of Engineering and Technology 34, https://doi.org/10.1007/978-981-15-1952-9_2

treatise "The Philosophy of Technology" (Encyclopedia of All Things, 1973, pp. 114–120). In the 1920s, the term "technology studies" appeared in Japan. Since approximately the 1940s and continuing to the present day, a debate over technology studies with a focus on the essence of technology has existed in Japan, as represented by Aikawa Ki and Taketani Mio. Japan's "technology studies" are quite consistent with the Western philosophy of technology. In a Japanese encyclopedia, the Western philosophy of technology is introduced under the entry of technology studies. Some Japanese philosophical dictionaries indicate that their interpretation of technology studies is identical to the term "philosophy of technology" in German (Philosophical Dictionary, 1981, pp. 303–304). In the past thirty years, many scholars have revealed greater interest in research into the philosophy of technology. A conference on philosophy of technology was held in the United States for the first time (Carpenter, 1980, pp. 40–46), and a number of papers and monographs had been published (Rapp, 1974).

Along with the emergence of the Institute of Philosophy of Technology and a group of philosophers of technology, there has been new progress in the research of philosophy of technology in Germany (Bugliarello & Doner, 1973). Scholars in Japan including Akihiro Hoshino, Seiji Nakamura, Satoru Hiroshi, Murata Miyabeiro and so on have completed many works on technological issues. (Yuan, 1981, pp. 232–235) So far, we have not seen any references to the terms "philosophy of technology" or "technology studies" by the academic community of the Soviet Union, but in fact there are quite a few works that explore this topic under different names (Hideo Nakasu, 1982).

Although the Western philosophy of technology and Japanese technology studies hold different views in a variety of matters, both of them attach great importance to overall research into technology and its laws of development as well as the general issues that are related to technological development. Murata Fujiro, a Japanese scholar, states that technology is entitled to two missions: "technology studies should be responsible for exchanging ideas among various branches of technology, and also responsible for putting technology into the world as a system." (Dialectic of Nature Group in Northeast Institute of Technology, 1981, pp. 43–44). The issues discussed in academia of Western philosophy of technology consist of the concept of technology, the driving forces behind technological advancement, the relationship between technological revolution and technological progress, the status and role of technology in social systems, the principles and values of technological history, and so on.

We believe that it is necessary to carry out the study of philosophy of technology under the guidance of Marxism, as is appropriate for China's status quo. Not only should China's philosophy look into the general laws of nature self and human cognition to nature, and the general laws of development of science, but also it should be concerned about the general laws of remaking nature and the general laws of development of technology. Basically, our two major tasks are directed at understanding and remaking nature. In some sense, remaking nature is more important than understanding it. The technology for remaking nature should be based on the scientific understanding of nature. Meanwhile, the technology used to remaking nature has its unique regularity. The philosophy of technology is concerned with the knowledge of the general laws of remaking nature. If philosophy also has its branches, it should not only include philosophy of nature, philosophy of science, and philosophy of history, but also philosophy of technology.

The four modernizations of socialism that we aim to achieve include scientific and technological modernization. Science and technology are closely related to each other but also exhibit differences (Chen, 1982). The modernization of technology is of special significance to the prosperity of society and the economy. However, not all kinds of technology are acceptable, nor are all measures conducive to the development of technology, as we must consider many concrete problems and fundamental principles along the way. To formulate and implement fitting technological strategies and policies in the proper way, we can not only rely on physics or chemistry, nor metallurgy or electrical machinery alone. We must understand the universal laws of technological invention. In other words, in order to develop engineering and technology, it is not enough to focus on manpower and material resources or specific technological methods; we should ensure it has a theoretical basis to solve such problems as the direction of technology, the driving force of technological development and the technological method, etc. By doing so, our technological career will be more congruous with economic and social development, the technological system will be more reasonable, and the technological measures will be more effective. The philosophy of technology should be the theoretical foundation for the development of engineering and technology.

The philosophy of technology is not only relevant for working philosophers, but also for working technicians. As Tani Mio once said,correct technology studies should be a powerful "guiding principle" for technological experts to develop technology. Obviously, this can be realized only when technicians and philosophers of technology are both concerned about the philosophy of technology. At the same time, the research into philosophy of technology can also be of use to management professionals, social workers, and history researchers, as is precisely Murata Fujiro's point when he says "making technology studies a part of sociology that people who are not technicians will also need" (Takenori Wood, 1981)

In a word, the philosophy of technology should be the philosophy of human to remake nature, the philosophy which guides human's remaking nature, the guideline for the development of engineering and technology, as well as the directing principle for the achievement of modernization of science and technology. We should research and establish a philosophy of technology that exhibits scientific and Chinese characteristics while adopting theories from Western philosophy of technology and Japanese technology studies for reference.

2.2 The Practical Significance of Studying the Philosophy of Technology

It is arguable whether it is necessary or not to call the research of general laws of remaking the nature as philosophy of technology, or to call the area of exploring general issues of engineering and technology as technology studies. More worthy of discussion is the question of whether or not such scholarship or areas of expertise can be established with practical value. We do not believe that the name for a subject abroad should be necessarily identical to the name used in China. Nor do we suggest copying a foreign philosophy of technology or technology studies, exaggerating its status, and regarding it as China's philosophy of technology. Western philosophy of technology and Japanese technology studies can be taken as references for their beneficial information and reasonable ideas, some of which attempt to explore problems from the aspect of historical materialism. At the same time, we should note that whether we talk about philosophy of technology from Germany or the United States, or Japanese technology studies, we must always realize that they are in the process of forming and do not exert a great influence on engineering and academia. Moreover, foreign philosophy of technology (or technology studies) is not directed at the problems faced in China after all. Thus, its way of asking and answering questions in part needs to be analyzed and criticized.

We believe that whether philosophy of technology can be established and whether it is necessary to conduct research into the philosophy of technology depends on real needs. In the course of advancing the four modernizations, especially for achieving modernization of science and technology, there are a number of practical and theoretical problems that are worth studying. These problems are not handled by existing disciplines, but belong to the philosophy of technology. Conversely, these studies can only be viewed as a development within the framework of foreign academia at most.

Many practical problems need to be investigated by philosophy of technology, for example:

Firstly, the relationship between understanding and remaking of nature, the relationship between science and technology, the role and significance of technology in economic and social development, and the question of the status of technology in the bidirectional relation between "science-technology-production" and "production-technology-science".

Secondly, we should investigate the restraints that economic and social factors place on technological advancement, as well the influence of social systems and social incentives on technology, and the importance of technological strategy.

Thirdly, we should research the differences and connections between the standards of technological progressiveness and the standards of applicability, as well as the relationship between technological invention and technological improvement, and finally the principles of technological evaluation and technological prediction.

Fourthly, we must explore the internal relationship of technological development, thought and practice of technology, technological structure and its rationalization, the status, role and interrelationship among each technological factor, cutting-edge technology and "traditional technology," and also military technology and civil technology.

Fifthly, we can apply the philosophy of technology to probe technological research systems, including the relationship between basic research, applied research and development research, the relationship between technological possibility and technological reliability, and the relationship between technological research and development.

Sixthly, we can address the conditions of technological development, including the relationship between independent research and technology importation, the relationship between technology development and technology promotion, and the relationship between technological competition and technological cooperation.

Seventhly, we can study the methodology of research in technological science and engineering technology, including the relationship between invention and discovery, the role of theory and experience in technological development, scientific experiments and technological tests, and general principles of engineering design.

Eighthly, we can use philosophy of technology to achieve coordination among technology, nature and humans, address human factors in technological activities, as well as the social status and activities of engineering and technical personnel, etc.

Two questions naturally arise after listing these issues: should the research into such specific issues be a philosophical task as well as a kind of philosophy? Is there any pressing practical value to investigate such abstract questions?

As for the highest level of philosophy, its concepts and propositions are highly generalized. They reflect on whether the origin of the world is material or consciousness, whether things in the world are static or moving, and inquire about the most common laws of nature, society and human beings. However, the problems mentioned above are relatively concrete. It is comprehensible to see why people do not call the study of these issues philosophy of technology but instead technology studies or another name. Even though it is not called philosophy of technology but rather technology studies, the content of these issues still has the nature of philosophy of technology. The content of philosophy should be understood not only from its highest level, but also from its lower levels. Dialectics of Nature explores the issues of nature, and its categories and laws do not always apply to social mentalities and the human psyche. However, Dialectics of Nature still belongs to a philosophical discipline. In other words, it is the natural philosophy of Marxism. Historical materialism examines the issues of social development, including the role of the geographical environment and population growth, and the relationship between economic foundations and superstructures, which are also not applicable to nature. But there is no doubt that historical materialism is an important part of philosophy, and that it can also be viewed as the historical philosophy of Marxism.

Although the problems discussed above do not occupy the highest level of philosophy, they are relevant to fundamental problems of human's remaking nature and general issues of technological development. Neither of these issues are research objects of natural science or technological science, nor do they belong to the domain of social science. These issues are not unique to a capitalist society or a socialist society. These are reasonable and sufficient grounds on which to term the study of these fundamental and universal issues the philosophy of technology, that is, a philosophy of man's remaking nature.

So, are these issues discussed above so abstract that they do not have a close link to real life? Is there any urgent need to study these impracticable issues for the cause of socialist construction? To be sure, the content of philosophy of technology is relatively "abstract". However, a philosophical principle does not lose concrete meaning due to its abstraction alone. Conversely, it is often difficult to appropriately deal with specific issues without a proper understanding of the questions of principle. The issues raised above will not be analyzed one by one, but one or two of them will be further explained.

The value of technology is a question of practical significance, and it should serve as one of the fundamental contents of philosophy of technology. In today's society, could the value of technology possibly be underestimated? Are there any well-educated scholars who do not know that scientific and technological work includes both science and technology, and that the modernization of science and technology consists of scientific modernization and technological modernization? In fact, this is not always apparent to everyone. Though some may fail to fully evaluate the natural sciences (especially basic sciences), the value of science has generally been acknowledged on the whole, which mainly can be seen from the assertion that "science is productivity" (which is also a philosophical proposition). Yet compared to an awareness of the importance of technology, this assertion is still not as commonly acknowledged. It has been reported that staff of Engineering and Technology Department of Chinese Academy of Sciences are less in number than that of other departments, that giving publicity to scientists overshadows the commending of the people actually working in the field of engineering and technology, that the proportion of outstanding young students studying science is higher than that of those who study engineering or agriculture, and that the manpower and resources invested in technological research are insufficient, so on and so forth. Due to the tendency neglecting technology, we must accordingly emphasize the enhancement of philosophical research into technological science and engineering technology (Liu, 1982).

The existence of practical problems to a large extent is attributed to the under-exploration of the theoretical value of technology. To solve this problem, we must explore the value of technology from three perspectives.

One is the economic value of technology. It is widely known that science is productivity. The question is how to view this in concrete terms. In other words, it is necessary to deeply study how science can be transformed into productivity. It could be explicitly stated that only through technology can science manifest as realistic productivity. This should be a fundamental principle of philosophy of technology, reflecting how the idea can be transformed into material force, and how scientific theories can guide production practices. We believe that not all natural science theories can yield productivity (undoubtedly, the value of science should not be evaluated from the aspect of whether it can lead to production benefits alone). Only when basic sciences can be transformed into technological science and engineering technology will they contribute to economic prosperity and the strength of a country. In this sense, the formula of "science-technology-production" is correct. Obviously, not only do the sciences mentioned here refer to basic sciences, but also include technological sciences and applied sciences. It is assumed that if the notion "only through technology can science turn into a productive force" can be thoroughly accepted, such problems as insufficient manpower and material resources of technology research, low social status of engineering technicians and other specific problems related to the development of technologies will be relatively easy to solve.

Secondly, the social value of technology is an important aspect also. It is not enough to measure technology in terms of the economy. The development of technologies such as military technology, medical technology and environmental protection technology can not directly lead to an increase in social material wealth, but take economic consumption as their prerequisite. However, these technologies have great social significance. Although state power and the quality of social life are closely related to economic development, they also depend on technological conditions with a wide range of social values.

Thirdly, the scientific value of technology must be considered. Science has technological value, and human knowledge and theories can be transformed into applied technology. What should not be neglected is that technology has its own scientific value, and this drives the advancement of science. Science has clearly played a guiding role in producing technology today. Meanwhile, the relationship of "production-technology-science" still exists. Production requirements are directed at technological reform and advocating promotion of technological development. Technology promotes the development of science also. This situation can be understood not only in terms of the progress of modern technological science and applied science, but also can be seen from basic science. Modern high-energy physics, astrophysics and so on are relatively independent of direct production. Yet, without technologies such as large-scale accelerators and radio telescopes, these disciplines would be difficult to move forward. The relative independence of these basic disciplines from production technology is strengthened, and their relative dependence on production technology is also strengthened. The scientific value of technology is obvious from the development of technological science and applied science. The research into technological science and applied science should be carried out thoroughly, and much attention should be paid to the development of engineering technology at the same time. Both of these aspects are closely connected.

The unique value of technology must be understood from the point of view of technological characteristics. It is of real significance to discuss the characteristics of technology, and such characteristics should be involved in the exploration of philosophy of technology. Generally speaking, technology has practicality originally. No matter how important the role of scientific theories plays in technology, technology always includes of practical and empirical factors. If metallurgical theories in books can be viewed as alloy processing technology, or electronics and mathematical theories can be viewed as electronic computer technology, there is no need to keep secrets of alloy processing technology and electronic computer technology. And if the establishment of scientific institutes can solve all technological problems, there is no need to retain these technical institutes. Additionally, the economic benefits of technology should also be given full attention. Upon evaluating a scientific achievement, people pay more attention to its correctness without too much concern about the benefits it generates. In this sense, basic natural science is viewed as "pure science". However, technology cannot focus on reasonable principles alone. In other words, it cannot be so "pure". It always takes its economic and social benefits into consideration. When capitalists adopt a new technology, it is not only because of its advancement or possibility in principle, but also because of its potential for generating profit. Therefore, capitalists are not willing to make profitable technology open or transfer its possession to others. Due to practicality and economic value, technology also has its confidentiality. If scientific principles are universal and international, then the scientific knowledge communication is less restricted. Science can be viewed as a "gift", while technology is something that can and should be kept confidential. To a larger extent, technology can be "traded". Furthermore, the application of technology is inseparable from the supply of raw materials, resources, energy and other conditions. The conditions in various countries and regions are different, which gives technology its own regional characteristics.

On the basis of the above discussions, it is necessary to give special significance to technological research and development of technology. A country should not be without its own scientific research, nor can it not have its own technological research. If the characteristics of technology are not fully realized, then we are likely to make mistakes in practical work.

In order to develop engineering technology, we must first solve many theoretical issues. For example, suppose one person makes an unprecedented technological creation, and its technological advancement surpasses the world standard. Meanwhile, another person makes minor technological changes to the existing equipment or technology with greater economic benefits in application. How should these two persons be rewarded? This is a practical problem, and the solution to it depends on the theoretical issues of technological evaluation. The theoretical research of technological evaluation standards has a fundamental practical significance. Technological inventions are based on the finding of new possibilities of technology. A technological study usually explores what kind of goal one technological measure can achieve. However, in terms of technological application, it is essential to take technological reliability as the prerequisite, except for the investigation of the possibility for practice. To solve the practical problems of technological application, it is also necessary to specify the role of studying technological reliability, as well as the system and policies for solving technical reliability research. All of these are inseparable from the study of technology theories. Methods of researching engineering technology and its inventions are different from that of scientific research, as well as its methods of scientific discovery. Philosophy of technology should look into the methodology of technology.

Technology is a way for human beings to remake nature. Wherever there is man, there is technology. At this point, technology is a social phenomenon. The research for technology should not be conducted from analysis perspective of disciplines only, but we should instead look into its links to other social phenomena as well as its evolution and development in an integral way. Whether it is called philosophy of technology, technology studies, or technology introduction, this kind of research is ultimately necessary and should be beneficial to technological development.

In the writings of Marx, Engels, and Lenin, there are many discussions on the fundamental issues of technological development as well as surveys and evaluations of ancient technology, steam technology and electric power technology. However, in the 19th century and early 20th century, engineering was far below today's level and without the same importance assigned to technology today. Moreover, it is impossible to give a more detailed analysis of the regularity of technological development when the task of socialist construction was not on the agenda. A serious investigation of technological universality and exploration of the rules of remake nature require our greatest efforts.

2.3 On the Development of Philosophy of Technology as a Discipline

As a relatively independent field of knowledge and an academy or branch discipline with special significance, philosophy of technology has realistic and necessary reasons for existing. In addition, we must observe what kind of knowledge systems or disciplines it contains or whether it has a unique set of concepts and propositions. One discipline must include and adopt related concepts from other disciplines, yet it cannot be established if the concepts it adopts and the content it involves are developed and defined completely by other disciplines.

We believe that philosophy of technology exhibits its own research concepts, such as "technology" "technique" "skill" "artistry" "technological revolution" "technological supremacist", and "technological pessimism". Engineering technology experts are dedicated to the development of various special technologies, whereas philosophy of technology needs to investigate the general concepts of technology, the differences and connections between it and techniques or skills, as well as the structures of technology and the factors that influence them. The exploration of these problems should require the participation of engineering experts. By participating, they have stepped into the field of philosophy of technology.

In terms of the research framework of technological philosophy, some tentative ideas are proposed from the draft by Professor Shangang Zou (邹珊刚) from Huazhong University of Science and Technology, along with my discussion with Professor Shixu Guan from Harbin Institute of Technology, and my final supplement and arrangement of it. The release of research plans is an attempt to illustrate which concepts philosophy of technology investigates. More importantly, we aim to seek advices to make this research framework more complete.

Firstly, it is necessary to study the Marxist theories and views about technology and take them as the guideline for the study of philosophy of technology. In

Capital, *The Use of Machines• Natural Forces and Science*, *1844 Economics-Philosophy Manuscripts* and other works, Karl Marx gave profound elaborations about the structure of technology and the development and signification of technology, which should be well-interpreted with the development of modern technology.

Secondly, the main works on philosophy of technology and technology studies abroad should be translated and introduced to China as references.

Thirdly, the status quo of China's technological development should be well-studied with a deep understanding of practical problems, experiences and lessons in the process of making and implementing technological policies. This should be followed by an analysis, summary, and advice for future development. Meanwhile, technical experts should be invited to participate in this field.

Fourthly, research into technological history should be carried out. Technological history is an independent discipline that is not equal to philosophy of technology. Just as natural history is not philosophy of nature, science history is not philosophy of science, and social history is not historical philosophy. However, the history of technology is the foundation and logical basis of the study of philosophy of technology. Without the research of technological history, philosophy of technology will fall into conceptual arguments or become empty talk. Simultaneously, the study of technology history also needs philosophy of technology. Some technological historians are also philosophers of technology. As for the ideological aspects of technological history or the history of technological thought, they can be taken as the content of philosophy of technology just as the history of scientific thought can serve as the content of philosophy of science.

We believe that if this work can be completed effectively, the study of philosophy of technology will yield many achievements and be beneficial to the construction of socialism and the advancement of academic development.

References

Bugliarello, G., & Doner, B. (eds) (1973). *The history and philosophy of technology.* Library of Congress Cataloging in Publication Data.

Carpenter, S. R. (1980). *Advances in American philosophy of technology* (vol. 5, pp. 40–46). Beijing: Science and Philosophy (edited by the Association for the Journal of Dialectics of Nature, Chinese Academy of Science).

Chen, C. (1982). Unification and differences between science and technology. *Guangming Daily,* 10, 15.

Dialectic of Nature Group in Northeast Institute of Technology (ed.) (1981). Murata Fujiro's technology studies. *Research Materials on the Structure of Science and Technology*, *2*, 43–44.

Encyclopedia of All Things. (1973). *VII* (pp. 114–120). Tokyo: Shogakukan Co., Ltd.

Hideo Nakasu (1982). *The Soviet Union's technology studies, translated from the history of science*, vol. 1.

Liu, Y. (1982). Developing China's scientific and technological work. *People's Daily, 6,* 2.

Philosophical Dictionary (1981). Tokyo: Heibonsha Co. Ltd., 303–304.

Rapp, F. (ed.) (1974). *Contributions to a philosophy of technology*, D.Reidel Pub. Co.

Takenori Wood (1981). *Kenji Nakamura's technology studies entry*. Beijing: Science and Philosophy, Ibid, No. 5:222–225.
Yuan, D. (1981). *Science and technology research bibliography, abstracts from Japanese journal technology and humanity*. Beijing: Science and Philosophy, Ibid no. 5, 232–235.

Chapter 3
The Features and Values of Changshu Chen's Philosophy of Technology Ideas

Qian Wang

Abstract From the perspective of international study of philosophy of technology, Changshu Chen's ideas of philosophy of technology as a system of ideas with unique features, influenced by Marxist practical materialism, having taken in some relative ideological elements from modern Western philosophy of technology and meanwhile based on philosophical reflection upon the experience of the technological development in China. Differed from the paradigm of analyzing technological activities in terms of concepts and structures, his ideas attached great importance to analyzing the function and value of technological activities, to being problem-oriented, to the organic relationship and systematic study of the relationships between relative elements of technological activities. In his ideas it could be found modes and ways of functional and value analysis aimed at the practical situation of technological development in China, and a summary of the laws of technological development from the comparative perspective between Chinese and Western. His ideas will not only be helpful in ensuring Chinese engineers' and technicians' philosophical attainments but also in making the features of Chinese philosophy of technology known by the corresponding international realm so as to promote deep spiritual communication.

Keywords Changshu Chen · Philosophy of technology · Function · Value

Professor Changshu Chen (1932–2011) is a pioneer in modern research on philosophy of technology in China. His ideas of philosophy of technology has not only exerted a great influence on the direction and methods of modern Chinese research in philosophy of technology, but also on international academic exchange, and it has become a significant reference in the understanding the distinguishing features of Chinese philosophy of technology. Regarded from the perspective of international research on philosophy of technology, the distinguishing features and values

Q. Wang (✉)
Department of Philosophy, Dalian University of Technology, No.2, Linggong Road, Ganjingzi District, Dalian, China
e-mail: qianwang@dlut.edu.cn

Q. Wang (ed.), *Chinese Philosophy of Technology*, Philosophy of Engineering and Technology 34, https://doi.org/10.1007/978-981-15-1952-9_3

of Changshu Chen's ideas are of great theoretical and practical significance. It is helpful to inherit and carry forward this ideological heritage, as well as promoted the intensive study of philosophy of technology in China and the academic exchange of ideas worldwide.

3.1 Distinguishing Features of Professor Chen's Philosophy of Technology

Changshu Chen's ideas of philosophy of technology were formed in the background of evolution of modern Chinese philosophy, with its unique theoretical characteristics and methodological features of localization in China. Professor Chen's ideas of philosophy of technology were characterized by the following three aspects.

To begin with, it stresses the importance of function and value analysis of technological activities, although that doesn't mean excluding the conceptual and structural analysis concerning the essence of technology. As a result, his study on artificial nature, on the process of technological activities and on the interactive relationship between technology and society is primarily done by means of function and value analysis.

It is noteworthy that, in his book *Introduction to Philosophy of Technology*, he intentionally did not bring up the definition of technology until the first section of Chapter 5, "*Technology and Its System*". This arrangement has profound implications. "The definition of technology is not essential or critical to an engineer," he said. "Most engineers don't need to depend on the definition of technology in order to do their work well and excel in their field" (Chen, 1999: 91).

In philosophical research, deep exploration of already familiar concepts may give rise to a new understanding, provided that those concepts are put in the context a new era or a wider perspective so as to reveal new characteristics. A few examples are understandings of Norbert Wiener's "purpose", Heidegger's "being" and Bruno Latour's "actor". However, when carrying out practical activities it is necessary to tacitly approve of the existing understanding in order to think efficiently before new characteristics or questions about those concepts are found. Nowadays, most engineers undertake technological activities with the common and habitual understanding of "technology" in mind, as philosophers' logical analysis of the concept of "technology" hasn't exerted any new substantial influence on engineers' practical activities.

In comparison, the function and value analysis of technology, especially the study of ethics of technology, has become more and more influential. As function analysis touches upon the design and use of technology (Houkes & Vermaas, 2015), and value analysis relates to the ethical aspects of technology as well as its impact on society and natural environment, function and value analysis should be the key research area of contemporary philosophy of technology. This distinguishing feature of Professor Chen's ideas of philosophy of technology is shared by Professor Deyu Yuan and his team to a great extent, has guided many Chinese scholars into the research of the function, value, ethics and social impact of technology. Some different or even contradictory ideas should be mentioned at this point.

In 2011, Professor Huaxia Zhang and Professor Zhilin Zhang pointed out in their article that the central issue of philosophy of technology is the epistemology and logic reasoning of technology. They call on a change in the direction of philosophy of technology towards the study of theory of knowledge and technical logic (Zhang and Zhang, 2001). Professor Changshu Chen and Professor Deyu Yuan subsequently argued that the central issue in the study of philosophy of technology should be significantly different from that of philosophy of science, and that axiology of technology should be the center of the contemporary study of philosophy of technology (Chen and Yuan, 2001).

To some extent, these two articles reflect the authors' different views. The former advocates the application of some of the research approaches and methods of philosophy of science on the research of philosophy of technology, while the latter emphasizes the difference between technology and science in terms of value and social impact, as well as in terms of the actual need of technological practice and engineers. These two approaches emphasize different aspects in of the research on philosophy of technology, but they are not mutually exclusive.

In 2013, Professor Guolin Wu (吴国林), in his paper "On the Research Guideline of the Philosophy of Technology", agrees with Professor Huaxia Zhang and Zhilin Zhang, and criticizes the views of Professor Chen and Professor Yuan, as he asserts that only the study of epistemology and reasoning logic of technology by means of logical analysis can be counted as philosophical research (Wu, 2013). This view regards the research on the function, value, ethics, and social influence of technology as a non-mainstream or even non-orthodox type of research on philosophy of technology. This judgment is worth revisiting. Can the modes and methods used in the study research on philosophy of science fully become the standard in assessing the maturity and direction of the development of the research on philosophy of technology? Are the research to epistemology and reasoning logic of technology necessary prerequisites for the research on the function, value, ethics and social impact of technology? Is it proper to say that the conceptual and reasoning analysis of technology can be considered as philosophy, but the function and value analysis cannot? From the point of view of the development of philosophy of technology in China and abroad in recent years, it's not difficult to give objective answers to these questions.

This characteristic in Professor Chen's ideas of philosophy of technology has its profound theoretical basis and practical foundation. Influenced by Marxist practical materialism, Professor Chen emphasizes technology should be understood mainly as a practical activity, not merely as a form of knowledge. A philosophy of technology that takes "practice" as its logical starting point has to be closely connected with the function and value analysis of technology. Such understanding is related to his personal practical experience.

In college, he majored in mining engineering and worked as laboratory technician in the Electric Traction Automatic Control System Laboratory, in the Northeastern University, engaging in all kinds of professional technological work. This technology background and his first-hand experience in modern technology have influenced deeply his research on philosophy of technology (Chen and Wang, 2015). Professor Chen repeatedly stressed that philosophy of technology is a philosophy for engineers, that speaks for engineers, and that establishes a dialogue with engineers (Chen, 1999: 3). Consequently, the philosophical problems that concern engineers the most should be the key problems to be solved by the philosophers of technology.

Secondly, it is problem-oriented. Professor Chen has posed 35 problems concerning the research on philosophy of technology which had a great impact in the field (Chen, 2002: 122–133), and has repeatedly stressed "problem consciousness" on various scientific and academic exchanges. This distinguishing feature has a lot to do with the characteristics of technological activities and those of the development of modern Chinese technology.

One basic difference between technological practice and scientific research is that the former cannot go too far in idealizing the research object, process and theoretical models. Technological practice must consider numerous real-life social factors, including natural environment, social demands, economic profit and work efficiency. Therefore it is difficult hard to build an ideal model for technological practice that will meet strict demands in abstract reasoning. Correspondingly, unlike with philosophy of science, in philosophy of technology it's not easy to establish a research model or code that meets these kinds of demand.

A feasible research approach is to start trying to solve specific problems, continuous summary all the philosophical issues that may arise during the technological practice and abstract them to a theoretical level, thus continuously perfecting the research framework of technology ontology, epistemology, methodology and axiology. Also, the ability to interpret and solve real problems should be taken into account in judging the research results of philosophy of technology. When discussing how to maintain the vitality of the research on philosophy of technology, Professor Chen stressed: there is no status without distinguishing features, no competence without foundation also, and no future without application (Chen, 2002: 106). His special emphasis on the application of the research on philosophy of technology is closely related to his problem-oriented approach.

This characteristic problem-oriented approach to research is closely related to the particular circumstances of the development of modern Chinese technology. Modern Chinese technology has developed while trying to catch up with advanced technologies from developed countries, with new manifestations of the conflicts between modernization and localization constantly emerging. The modernization of Chinese technology has its own new features in dealing with environmental and social impact problems as well as problems derived from cultural notions. As it's nearly impossible to borrow some ready-made theoretical basis and methods from the Western countries, Chinese scholars need to explore deeper into reality and identify the problems themselves so as to create their own research characteristics.

In Professor Chen's writings on philosophy of technology, he discussed many technical problems and cases from real life extracted from modern technological practices in China, reflecting his concern about China's major social problems and his strong sense of social responsibility.

Thirdly, it attaches importance to the organic relationship between the different elements involved in technological activities and its systematic study. A great part of Professor Chen's philosophy of technology discusses this particular aspect, including the relationship between intellectual and non-intellectual factors, between human and material factors, between technology and production, between technology and engineering, between technology and various kinds of social factors, etc. When analyzing these organic relationships, he effectively combines dialectical and systematic thinking, keeping a flexibility of mind while criticizing current malpractices and producing actual effectiveness.

For example, he pointed out that the subordinate relation of importance among the different elements in a technical system cannot be decided on the grounds of whether they are necessary or expendable (Chen, 1999: 104); the decisive effect of technology on society cannot be regarded as "sole" or "one-way" (Chen, 1999: 190); the discourse cannot be limited to the harmony between man and the environment and ignore the necessity to transform the natural environment (Chen, 1999: 269), etc. Carefully disentangling such close concepts not only requires an extraordinary ability in logical thinking but also the resourcefulness of a dialectic thinking, as well as careful and systematic reflection.

This feature of Professor Chen's thought stems from his profound understanding of the nature and function of philosophy. To him, philosophy is more than a system of knowledge, it is rather a first-hand experience, the ability to use the intellect to see through problems and acquire an enlightening and penetrating insight. This feature is clearly reflected in his academic exchange, philosophy teaching and student tutoring (Wang, 2009). The analysis of the organic relation among the relevant elements of technical activities cannot stop at the simple patterns of dialectical relationship, i.e. "it is… as well as…"; conversely, it must penetrate into the actual process of technical practice, and reveal where misconceptions about the issue easily arise, in order to single out a thinking approach which corresponds with the operational and practical characteristics of technical systems.

3.2 Values of Professor Chen's Philosophy of Technology

The distinguishing features of Professor Chen's philosophy have determined its theoretical and practical value. From the perspective of international research on philosophy of technology, or rather, compared with other schools of philosophy of technology in China and abroad, the values of his philosophy can be summarized as follows.

To start with, it provides models and methods for function and value analysis aimed at the actual situation of the development of Chinese technology. His philosophy, rooted in modern Chinese technological practice and capable of pertinent function and value analysis, has established a theoretical analysis model relevant to the actual condition of the development of Chinese technology. Its main points are:

1. It integrates modern features in its analysis of the natural and social attributes of artificial nature (artificial things) (these features include the application of social regulations, the attainment of support from society, assessed socialization, development pace, etc.), which can be equated to providing the background and scope for the theoretical analysis.
2. It analyzes the common problem of different technological areas (like policies, processes, control and guaranty), which touches on the requirement of multiple indicators, such as quantity, variety, quality, standard, knowledge, risk, profit, environment, etc., and reveals the relationship between different elements of technological practice. This is the foundation for the function analysis of various types of technology.
3. It analyses of the relationship between the material and intellectual factors of each technology, between the industrial fundamental technology and the industrial supporting technology, and between traditional and high technology, in order to form a network of relationships among technological elements, so as to lay the foundation for further value analysis.
4. It analyses the relationship between the origin of technology and social demands, of the shift from intellectual technology to realistic technology, and of the relationship between technology, production and engineering activities, so as to reveal the role played by the value factor in technological activities.
5. It analyzes the role of technology in society, including its influence upon economy, social life, politics and culture and, especially, on ethics and morality. This analysis is different from the sociological analysis of technology, as it is focused on the value aspect of technology and pays special attention to the change of thinking patterns.
6. It analyzes the constraint on technology imposed by society in terms of economy, politics, military affairs, culture and so on. As before, this analysis focuses on the impact of social factors on the value aspect of technology, and emphasizes the influence of social factors on people's technological notions, decision-making modes and evaluation standards, so that value analysis will play its due role in influencing people's viewpoint on the value and the development patterns of technology.
7. On the basis of function and value analysis, it focuses on the development patterns and strategies of modern technology, especially on the relationship between "sustainability" and "development" in modern technology, and points out the internal conflicts between the two, their possible negative consequences and ways to conciliate these two aspects.

This model of theoretical analysis fully reveals the function and value attributes of various elements within modern technological practice and the relationship among them, thus helping people see clearly which function and value problems their own technological work will give rise to and how technological work and social life will interact with each other. In this way, it promotes the transformation of ideas, deepens the understanding of the nature of modern technology, strengthens the sense of social responsibility and reconciles conflicting views caused by modern technology.

This model of theoretical analysis suits the characteristic way of thinking of engineers and technicians, as they first need to use a macroscopic approach to get an grasp on the social attributes of artificial things, which they might not be familiar with, as well as on the common characteristics of different technological areas, and gradually incorporate it to their specific technological practical knowledge. Subsequently, from a microscopic point of view, they need to shape their own understanding. Finally, by gradually broadening their view, they need to hold this understanding and experience against the macroscopic social background, to get a thorough understanding of the function and value aspects of technology. This is in accordance with the engineers' and technicians' cognition development patterns.

Secondly, using a comparative outlook between Chinese and Western philosophy of technology, Prof. Chen summarizes the laws of technological development and puts forward the following 4 laws: "The self-increase of technology development", "The compromise and compatibility of technological construction", "The gradual transition of technological update" and "The cyclic prosperity and decline of technological evolution" (Chen, 1999: 137–158).

The first law absorbed some ideological components of Ellul's theory of technological automation (although limited and elaborated), specially as it points out the existing dependence and constraint between the reasonability and feasibility of technological goals and the improvement and effectiveness of technological means.

The second law has some traditional elements of Chinese culture, it mentions "the fish and the bear's paw, you can't have both at the same time" (meaning one can't have it all, a choice must be made), and also incorporates some ideas of the Japanese scholar Kititani Yutaka, although it differs from them in as much it emphasizes the need to choose a proper compatibility point and the parameters of a reasonable compromise.

The third law emphasizes cautiousness with the technological revolution and pays attention to the idiosyncrasy of the relationship between old and new in the midst of technological development. Technological update shouldn't skip necessary stages. At this point, the historical experience of Western revolution of technology is summarized, and the conflicts of ideas on China's modern technology development are further discussed.

The fourth law treats the characteristics of technological evolution from the perspective of function analysis and points out the inevitability of the birth and death, prosperity and decline of each specific technology. It particularly stresses the importance of avoiding the linear way of thinking in technological development. This not only shows the ideological characteristics of dialectical thinking, but also the influence of systematic thinking.

Professor Chen's summary of the four laws of technological development are based upon the actual condition of the development of modern Chinese technology, especially the ideological inclination which lays emphasis on social demands during the process of technological modernization, pursuing optimization, hoping for breakthroughs in development and advocating the complete replacement of old technology by new technology. These four laws have an enlightening impact on the technological modernization of other developing countries.

In the time of the Internet, high technology has been playing its guiding role in more and more realms and, as a result, related ideological conflicts are increasingly appearing. Further discussion of the theoretical and practical significance of the four laws is an issue worth exploring.

Thirdly, it is helpful to improve Chinese engineers' and technicians' philosophical cultivation. From the perspective of study and communication, Professor Chen's ideas of philosophy of technology are characterized by its accordance with the thinking characteristics of Chinese engineers and technicians. Using numerous real cases of technological practice to illustrate the principles and methods of philosophy of technology, it is able to explain profound theories in simple words.

He has analyzed and criticized the ideas of various modern Western schools of philosophy of technology, approving their reasonable parts and pointing out their flaws and limitations. Meanwhile, he compares them with traditional Chinese philosophy of technology to make them understand by Chinese engineers and technicians against the Chinese cultural background.

There is a problem worth careful consideration: how to help those devoted to philosophy of technology improve their technological background. If they do not have a basic understanding on modern technology and its development, they will not be able to explain in simple terms the philosophical problems found in typical cases of modern technological development, it will be difficult for them to communicate with technicians and, consequently, the efficacy of philosophy of technology will be compromised.

It's certain that scholars of philosophy of technology cannot be asked to master all aspect of contemporary technology, but it's possible for them to understand its basic characteristics and principles, and to accurately explain the technical issues related to philosophical thinking. Professor Chen himself sets a perfect example in this respect.

Although he has an engineering background, as well as extensive professional practice in technology, many of the cases analyzed in his writings involve knowledge beyond his expertise. This is achieved through the constant update and improvement of his knowledge structure and technology literacy. All though his life, he has held a keen interest in the history of technology and the latest achievements of modern technology, thus being able to maintain a high level of dialogue with technical experts of different fields. This is the ability that all scholars engaged in the study of philosophy of technology should acquire and master.

3.3 Conclusion and Prospect

From the perspective of international research of philosophy of technology, Professor Chen's ideas of philosophy of technology is influenced by Marxist practical materialism, while it has also absorbed some elements from modern Western modern philosophy of technology, and it's build upon the reflection on the development of modern technological development in China. In general terms, his philosophy belongs to the philosophy of technology of engineering tradition, although it is open to the humanistic tradition of philosophy of technology. His writings on philosophy of technology carry a valuable ideological heritage and will exert a long-term influence on the future development of Chinese philosophy of technology.

In the 1980s and 1990s, Professor Chen and his academic team paid close attention to serious practical problems existing in the development of technology, the technological innovation of enterprises and the engineers' ideological needs. This promoted the idea of communication between philosophers and engineers. Such research paradigm not only had a considerable impact on society at the time, but still is of great practical value at present.

The development of modern philosophy of technology needs to be rooted in modern Chinese technological practice and needs to form its own theoretical features. At the same time, an international perspective must be adopted in order to absorb the advanced ideas of foreign countries. Without a deep understanding of the progress of philosophy of technology in foreign countries, China's research on philosophy of technology will come to a standstill and become outdated. However, if it only concerns itself with the reviews of foreign philosophers and foreign ideological trends and loses its power to influence the practice of contemporary Chinese technology and the close contact with enterprises and engineers, philosophy of technology will lose its ground and become unable to fully achieve its function and value. In this respect, Professor Chen is an academic example worth following and his ideas are worth being constantly enriched and developed. This is the best way to commemorate him.

Modern Chinese technology and economy are rapidly developing. The development of Chinese technology has kept bringing about numerous philosophical problems waiting to be reflected upon. They need to be studied and solved by scholars of philosophy of technology. The vitality of philosophy of technology lies in the constant emergence of realistic problems arising from technology and influencing technology.

The development of modern high-technology, especially of nanotechnology, information technology, bio-technology and cognitive technology, have brought along considerable difficult problems in the areas of epistemology, methodology and ethics. In addition, risk, quality and environmental issues caused by conventional engineering activities may give rise to a number of conflicts of ideas concerning social responsibility of enterprises, professional ethics of engineers and social ethics, sustainable development, etc.

In order to improve Chinese engineers' capacity of technical innovation, especially their capacity of original innovation, it is necessary to integrate China's cultural background with the modernization process, and to carry out more research from the perspective of philosophy of technology. The research on philosophy of technology based on the reality of China's technological development is an academic pursue that Professor Chen devoted his life to, and deserves to be carried forward by those of us who have taken up the torch of this career.

References

Chen, C. (1999). *Introduction to philosophy of technology*. Beijing: Science Press.

Chen, C., & Yuan, D. (2001). Rethinking of the research program of philosophy of technology: Discussion with Prof. Zhang Huaxia & Prof. Zhang Zhilin. *Studies in Dialectics of Nature, 17* (7), 39–42.

Chen, C. (2002). *Changshu Chen's collected works on philosophy of technology*. Shenyang: Northeastern University Press.

Chen, C., & Wang, Q. (2015). Reflecting on the philosophy of technology in China in the form of five questions. *Technology in Society, 5*, 57–59.

Houkes, W., & Vermaas, P. E. (2015). *Technical functions: On the use and design of artefacts*. Dordrecht: Springer.

Wang, Q. (2009). *Philosophical uncovering*. Beijing: People's Publishing House.

Wu, G. (2013). On the research programs of the philosophy of technology. *Studies in Dialectics of Nature, 29*(6), 40–45.

Zhang, H., & Zhang, Z. (2001). The research program of philosophy of technology from the perspective of demarcation between science and technology. *Studies in Dialectics of Nature, 17* (2), 31–36.

Chapter 4
Technology and Its Value from the Perspective of Process Theory

Deyu Yuan

Abstract The perspective of process theory is a special research perspective adopted by modern Chinese philosophy of technology research. The analysis of a technological process includes two interrelated aspects: one is the diachronic analysis of technology and other is the synchronic analysis of technology, which demonstrates specific states of technology under certain temporal and spatial conditions. Technological value is generated through the process from purposiveness to natural function with purpose, then to effectiveness. How to keep the balance among production, waste, livelihood and life is a major issue in the development of technology and economy from the perspective of process theory.

Keywords Process theory · Philosophy of technology · Value · Production and waste

The perspective of process theory in the philosophy of technology research is a special research perspective adopted by modern Chinese philosophy of technology research, which distinguishing feature is to understand technology as process, the things transformed continuously, that is, technology presenting itself in different forms in different developmental phrases, which explains how there can be different but related definitions of technology. The perspective of process theory can be used to analysis the different form of technology and its values. Hereafter this research perspective will be the process theory of technology in short.

D. Yuan (✉)
Department of Philosophy, Northeastern University, No.195, Innovation Road, Hunnan District, Shenyang, China
e-mail: yuandeyu34@163.com

Q. Wang (ed.), *Chinese Philosophy of Technology*, Philosophy of Engineering and Technology 34, https://doi.org/10.1007/978-981-15-1952-9_4

4.1 The Raising of the Process Theory of Technology

The process theory of technology was officially proposed in 1981 when I presented my paper "Technology is a Process: Some Thoughts on the History and the Theory of Technology" during the Second Academic Conference on History of Technology in China. This paper firstly brought up the idea that technology is a dynamic process. Subsequently, I published other relevant papers, such as "Technology in History" and "Some Thoughts on the Fundamental Attributes of Technology". In the book *On Technology*, the dynamic process of technology was further demonstrated. In 1991, when the research on technological innovation and its related activities had developed widely in China, I submitted the paper "A Craftwork Study of Technology" at the National Symposium on Industrialization of Scientific and Technological Achievements in China, and proposed: "Over the past 5 years, the process theory of technology has been criticized by many; however, with the increasing attention on commercialization of scientific and technological achievements and more and more research on technological innovation being carried out, the process theory now seems to be acceptable" (Yuan, 2008). The reason why invention needs innovation is that invention is only the initial stage of technology. Only through innovation can invention be industrialized and transformed into industrial technology, and only at this final stage can the functions of technology be given fully play. This not only proves that technology exists in different forms during the process of transformation, but also that only in the process of dynamic transformation can we fully understand technology. Technology is a dynamic entity. As such, it can only be grasped in its process of transformation. When observed from a dynamic perspective, technology is a multi-factor system in transformation within a dynamic process. Therefore, technology manifests itself in different forms.

In the book *On Technology*, I tried to illustrate the domain of technology and the process of transformation by the following schematic diagram.

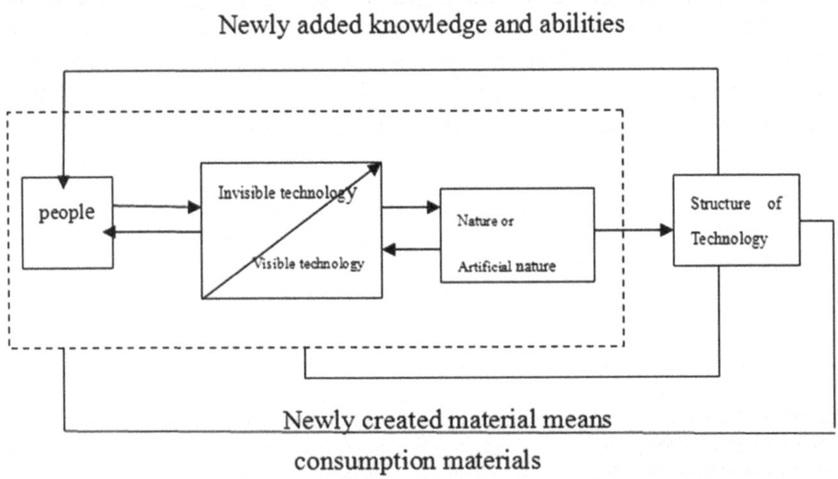

A schematic diagram of technology as a dynamic process

The area within the dotted line in the schematic diagram represents the existing domain of technology. As the media of human and nature, technology reflects human's effect on nature and vice versa, generation and development of technology are achieved in the reaction between human and nature. Human being's manipulating, controlling and transforming seems to be a kind of subjective initiative, but the fact is that human beings' technical responses are often passive, that is to say, it is often a "must do" response to a natural threat. Fear towards fire may be the original motive that urged primitive people to attempt to figure out how to control fire. Those anti-technological eco-protectionists are calling for people to love nature, but could the nature certain to love us? We have no idea how many lives are lost in earthquakes, volcanic eruptions, tsunamis, and avian influenza. We have to admit that immoderate use of technology surely caused today's ecological crisis and it did reflect the endless appetency and rapaciousness of human beings, nevertheless we still need to ask ourselves whether we could have a better life if inexistence of technology? Obviously not! When the development of technology causes environmental pollution and disables the basic living materials (such as fresh air and clean water) from meeting the needs of human beings, people passively face new technical problems. Then the so-called environmentally friendly technology will emerge. To solve environmental problems requires people to refrain from selfish desires, inflation and greed, and to do so either social mechanisms or technology are required. Therefore, the technology must be regarded as the intermediary between human and nature, and human's initiative to nature may be understood as a kind of "passive initiative", which is more consistent with the historical facts of the relationship between man and nature.

What human beings need is not only technology itself, but also technological results. Technological results and technology are two different things. Technological results have three functions: first, as a consumer's goods, to be consumed; second, as a newly created material resource, to become an element of technology and be added to technology's development process; third, as new knowledge or capability, to become the basis for the creation of new knowledge and capabilities. This new knowledge includes the reflection on the negative impact of technical results.

In the book *On Industrial Technology* (Yuan, Ding, & Ma, 2005), I defined technology from the perspective of the process theory as follows:

1. Technology is the unity of purposes and methods in a dynamic process.
2. Technology is the unity of human subjective factors and objective material factors in a dynamic process.
3. Technology is the unity of various forms of technology in a dynamic process.

This summary can be further abstracted into two sentences: technology exists in a dynamic process; technology is a kind of processional being. The former stresses the existence domain of technology, and the latter emphasizes its attributes, in order to differentiate technology as an entity, as knowledge and as an activity. The verb "exist" in the first sentence refers to being observed by practical philosophy, being felt and perceived by people. As Professor Feng Xiao said, “the sense of existence and existentiality are the human-centered concepts” (Xiao, 2005). When discussing the existence of technology, it’s better to stop at the experiential level of understanding, that is, it exists in as much we can experience it, as opposed to seeing it as a pure objective being. The existence of technology to which we refer has a structure. When trying to determine the characteristics of this existence, we need to determine the characteristics of its internal unity. What we call technology is the “unity during a dynamic process” and a “processional existence”, it’s precisely the conclusion of observing it from the described perspective.

Technology is a kind of activity unique to mankind. Any practical activity involves certain conflict between human beings and the outside world. The particularity of technical practice is mainly embodied in the conflict between man and nature, including the contradictions between purpose and means, subjectivity and objectivity, idea and reality and so on. Human subjective initiative and creativity are not only manifested in the comprehension and perception of the above contradictions, but also in the pursuit of the unity of contradictions. The so-called “unity in the dynamic process” is actually the process of seeking the unity of contradictions.

Contradictions between purpose and means are the basic ones in the relationship between man and nature. Setting up a technological purpose and seeking the means to accomplish it is the connecting thread that links together the activities of technological practice. The discovery of the natural laws that conformity with that purpose is the result of this search, it serves as the basic principle of technology and as a means of achieving purposes. In the process of technological practice, the means of achieving these purposes will vary as the goal changes. The purpose of technological invention, production technology and industrial technology is different, and the means to implement them are different as well. Moreover, there can be a variety of means to achieve the same end and the same method can achieve a variety of purposes, which causes the complex relationship between purpose and means. Therefore, the unity of the contradictions between purpose and means can often be approached in different ways, hence the multiplicity of solutions for technological problems.

The unity of the contradiction between means and purpose takes place under specific historical conditions. Social historicity is one of the basic characteristics of any practice, especially technological practice. The establishment of technological purposes and the search for means to realize the purposes must be based on certain social and historical conditions. The historical conditions include both subjective and objective conditions. Subjective conditions refer to knowledge, experience and skills possessed by an individual under specific historical conditions. Objective historical conditions mean objective material conditions, including materials, sources of energy, information, tools and equipment. Of course, people have unlimited imagination. Ancient people envisioned flying, controlling rainfall, freely transforming like the Monkey King, but these were only fantasies and desires, not real technological ideas. Under specific historical conditions, people’s rationality is limited, what reality can provide is limited, knowledge, experience skills and material conditions are all limited, therefore people's exploitation, control and transformation of nature are limited too. Heidegger used to resort to the concepts of "Gestell" and "herausfordern" to reveal the essence of modern technology, "gestell" means the use of something in a certain direction (i.e. take nitrogen from the air, take ore from land) to identify, fix or position one thing onto another. Therefore, gestell always means positioning or "forcing" (herausfordern) something into a certain unnatural status (Gunter, 1993). In fact, it is not only true for modern technology, but also for all kinds of technologies in history. Ancient people identified coal as fuel for cooking. People used coal for energy, providing power in the era of steam engines; people positioned coal as raw material in the chemical industry era. Each era has its special gestell and herausfordern, therefore, they are not only the essence of modern technology, but also the essence of all technology. This phenomenon can not only explain the social historicity and historical limitations of technological practice, but also shows that technological development is an endless historical process. The development of technology can only be a transition of herausfordern from one specific historical condition to another. This is the dialectics of technological development.

It also should be noted that some people classified technology as objective knowledge and believed "once technological theory or technological procedures are made public or externalized from the subject, they will exist independently. If so, how can we still talk about "subjective elements" "human factors"' or "subjective factors"? (Peng, Lin, & Li, 2002). This viewpoint is a mistake. According to the process theory, technology exists in a process, the knowledge that becomes a technological factor can only be knowledge acquired by people, so it is a subjective factor of technology. The knowledge in a book, or externalized knowledge can obviously exist independently, but neither can become technological factors if they are not grasped by inventors or innovators. Therefore, in order to promote invention and innovation of technology, the communication and integration of knowledge should be promoted, in order that individual knowledge becomes shared knowledge and tacit knowledge becomes explicit knowledge.

The contradiction between idea and reality is another basic contradiction in technological practice. From the viewpoint of epistemology, it is obvious that ideas appear after the related object or reality; but seeing technology as a human practice, ideas occur first, then they are turned into reality. As K. Marx said, the difference between how honeybees build their hives and how humans build their houses lies in the fact that the house must be conceptualized before it can be build. This is one characteristic of human practice, it should not account for an absolute understanding. Sometimes idea and reality influence each other in technological practice. "Practice is a dual process: on the one hand, the subject creates a demand for the object based on a "human criteria", and turns his "purposeful demand" into reality. This is called objectification of the subject (the object becomes the object demanded by the subject); On the other hand, the practical subject will regulate their own ideas and behavior based on a "material criteria" to carry out practical activities in accordance with the objective law, which is called subjectivization of the object. It is just in the process of unity of opposites between the objectification of the subject and the subjectivization of the object that people realize the unity of opposites between the transformation of the world and the transformation of themselves (Sun, 1998). From the practical point of view, the process of objectification of the subject and subjectivization of the object is a process of endless expansion. In this way, the technology of the conceptual form is transformed into the technology of materialized form, the technology of subjective form is transformed into the technology of objective form, and the technology of latent form is transformed into the technology of realistic form. It is also in this process that, in accordance with the "human criteria", technology accomplishes the objectification of the subject and is achieves human value, and in accordance with the "material criteria", technology achieves the subjectivization of the object and achieves material value, and also achieves its potential value. The unity of the contradiction between idea and reality is not only the transformation process of the technological form, but also the transformation process of the technological value. The scholars who support the process theory of technology must support both the theory of technological morphology and the theory of technological value.

After the introduction in China of Whitehead's organic philosophy and David R. Griffin's postmodern spirit, some people tried to incorporate my process theory of technology to the category of process philosophy, but the fact is that the dynamic process theory of technology is not a conclusion of process philosophy in the field of technology, but a philosophical generalization based on our own empirical analysis. The contradiction between the practical experience of industrial technology and the existing technological understanding has prompted me to put forward the process theory of technology. Technology is inherently in constant generation and transformation, and it is only in the process of generation and transformation that we can understand it in a comprehensive way. The existing definitions of technology often refer to the function of technology or the result of technology, but not to technology itself. The dynamic process theory of technology provides a new angle for people to study technology. This is its original intention.

4.2 Morphology Theory of Technology and Technological Innovation

I began to cooperate with Japanese scholars on a comparative study on the technological development between China and Japan in 1985. The initial idea was to provide references for technological reform carried out in Chinese enterprises at the time. However, we soon realized that the Japanese do not use the term "technological reform", what they use to describe similar technological improvements is "innovation" and that is why we changed the research title from "technological reform" to "technological innovation". The book *Comparison of Technological Innovation between China and Japan Enterprises* was published by Northeastern University Press in 1994 and this was the first time for the concept of technological innovation to be officially introduced in China. Subsequently, we visited a number of enterprises in Liaoning Province, China and handed out questionnaires about the status of technological innovation in local enterprises. All of this research allowed us to learn more about innovation in enterprises and also to fully validate my process theory of technology. The recognition of the process theory of technology inevitably leads to the recognition of the morphology theory of technology. I realized that technological innovation is the transformation process of technological form. We published a number of papers about this issue, and finally made a thorough but brief discussion in the book *On Industrial Technology*.

The analysis of a technological process includes two interrelated aspects: one is the diachronic analysis of technology, through which we can demonstrate the existence domain of technology and clarify the generation and transformation process of technology. This process is continuous, but within its continuity it also has phases. At different stages, technology assumes different forms. The other is the synchronic analysis of technology, which demonstrates specific states of technology under certain temporal and spatial conditions. These two aspects represent technology as the unity of external form and internal foundation. All matter has its roots and branches, and every event has its beginning and end. "Matter" refers to what can exist independently outside of human beings, while "event" refers to human behavior and activity. Technology is an object that includes aspect of matter and aspect of "event" which actually is the unity of matter and event. From the dynamic perspective, we can leave out neither matter nor event. Only in this way can we comprehensively understand technology, and also understand "object" in the context of the process unity of matter and event.

In technology, there are two kinds of relationship forming: one is the relationship between humankind and nature, the other is the relationship between human purpose and means. The contradiction between humankind and nature gives rise to technological problems and the need to transform nature. These technological problems embody the constraint that nature exerts on human beings and the predicament. Breaking free from this constraint and predicament is people's desire, and from it stems the objective pursuit of nature's transformation. Once there's purpose, the process of seeking the means of achieving it begins: the technological plan. The technological plan is not a subjective human conception. Different from artistic and literary conception, the technological plan must accord to the causality of nature and make use of it to realize the individual purpose of remarking nature. It's a cycle created by the action of thought on action, and the reaction of action on thought, only in this way can the basic principles of technology be discovered and technological invention take place. The basic principles of technology are the natural laws that conform to its purpose. People can use these natural laws to create artificial things in order to get rid of the aforementioned constraint and predicament. If nature's laws and causality are to be considered scientific principles, then it can be said that the natural laws that conform to the technological purpose can be considered technological principles.

The principle of technology is not meant for people to stop at the understanding of natural causality, but to guide people in the creation of artificial things. It is precisely from here that the extreme social constructivists of technology are in error, when they seem to think that people can arbitrarily construct technologies ignoring natural causality. They do not seem to admit that technological principles are discovered by people, as opposed to invent by people. Of course, this is by no means saying that technology is just a simple extension of science or a mere scientific application. The theory of scientific application of technology stresses in excess the idea that the fundamental principles of technology are but the result of scientific application, as if technology were merely a scientific discovery, not an invention. The invention in which natural laws that conform to a technological purpose depend on the discovery of the natural laws, but this does not mean that the invention of technological principles is merely the result of the research on natural science.

The theory of scientific application of technology provides many examples to demonstrate that technology comes from scientific applications. But we can also provide even more examples to show that many technological inventions, including modern technological inventions, do not come from scientific research. The first stone-knife made by human beings follows the wedge principle, but the invention of the stone-knife did not require the prior discovery of the principle; the first fire arrow made by human beings follows the principle of law of momentum conservation, but the invention of the fire arrow did not require the prior discovery of the law of momentum conservation. Maybe a sharp stone chip had scratched a human body, after which humans started using it to tear off meat from animals. The stone chip being able to tear off flesh was only the discovery of a phenomenon that is coincidental with the wedge principle. The Chinese rocket principle was discovered by the Chinese, but they did not understand the law of momentum conservation in the Song dynasty, they just discovered the potential of firecrackers, after which they could invent the "sky monkey", "high gun" and afterwards, the booster-fire arrow combination "fire dragon out of water" and the strap-on fire arrow "surefire of flying crow" etc. The discoveries we are talking about here are discoveries of natural phenomena. Under these circumstances the technical principles happen to coincide with the related scientific principles. Natural laws are discovered through scientific research, putting them in the service of a goal in order to form technological principles is an example of technological principles conforming to the related scientific principles. "Coincide" and "conform" are two different paths of cognition.

From technological conception, a technological principle is formed, which produces a technological idea. This is the beginning of technological creation. Although the process implies human action, at this stage this action only serves the technological conception. An idea is technology existing only in a human mind, subjective and intangible. Only through a process of objectification can the subjective technology change into a tangible technology. The objectification of the subjective technology can be expressed by symbols, graphics, words and material objects. With the objectification of subjective technology and the transformation of intangible technology to tangible technology, the purpose of technology will change into a visible and tangible technical function. Tangible technology is the real technology, as it creates something that previously did not exist in nature.

The objectification of subjective technology is the process of translating human thoughts into action. According to the idea conceived in people's minds and relying on their own experience, skills, knowledge and wisdom, through the use of tools the process is completed. Although there is a certain amount of design in this phase, it is not expressed with symbols, graphics, etc. It is similar to the way that artists do their job: after having a subjective model, a sculptor will also go through the above process to form a realistic sculpture. The products made in this way are always unique. A craftsman cannot produce exactly the same tools even if he repeated the exact same procedure.

A technician with technological conceptions represents their ideas in a symbolic and graphic manner, which is what we usually call technical design. The design process is the process of selecting the various elements that may contribute to the technology. The technical design used for manufacture might be in great part devoted to the tools and equipment involved. The goal of technical design is manufacture. But design at the invention stage (product design) is different from that at the engineering stage (engineering design). Although they both have manufacture as goal, product design is used to reveal the effectiveness of the invention. The artifacts produced in this stage can be called samples, belonging to the technical design of single products. Engineering design is for the manufacture of the product, belonging to the design of mass production manufacturing.

Technological invention is the process of choosing and creating means and methods for realizing purposes. The purpose of technology is embodied in the effectiveness of means. To achieve the same purpose, we may face a variety of choices of means, and the same technical principle may lead to a variety of technical inventions. As Changshu Chen pointed out in the article "On the Transformation from Science to Technology from a Philosophical Point of View" (Chen, 1994), "the transformation from science into technology is the transformation from the causality to the purposiveness, from the truthful standard to the technical utility, from the general rule to the diversification of technology". The efficacy of various inventions often determines whether they can be transformed into production technology and industrial technology.

Technological invention only provides the possibility of achieving a technological purpose. Although technological inventions are already objectivized technology, in terms of their social role, they are still in the potential stage. To turn the possibility of technology into reality, or to turn potential technology into practical technology, it is necessary to meet certain social and economic requirements. Only when the economic and social conditions allow, can an invention go on to be transformed into production technology and industrial technology. Socio-economic feasibility is actually the embodiment of technical purpose and efficacy. If the invention is efficient but cannot meet the socio-economic requirements, then this efficacy can only remain in the potential stage. Socio-economic feasibility is the decisive condition for the transformation of potential technology into practical technology. Socio-economic feasibility includes both social viability and economic viability. If they are both in accordance, technological inventions can quickly be turned into productive technology and industrial technology. When economic and social viability are not in accordance, two different situations may arise: first, sacrificing economic feasibility in order to meet social feasibility, as it is often the case with military technologies; second, sacrificing social viability to meet economic viability, as some entrepreneurs do in order to achieve the maximum profit. That is to say, as long as the conditions of economic or social viability can be satisfied, the transformation of technological invention into production technology and industrial technology can be realized.

The process of transformation from technological invention to industrial technology is known as technological innovation. Innovation was originally an economic concept proposed by American-Austrian economist Joseph. A. Schumpeter. The concept refers to the reorganization of production factors, including technological innovation and organizational innovation. The new Schumpeterian school believes that the central task of innovation is technological innovation, while organization and system are in the service of technological innovation. Conversely, the Institutional school thinks that the central task of innovation is system innovation, which should be served by technological innovation. However, both schools of thought regard innovation as a technological process entering the economy, thus affecting the economic growth and the periodicity of economic development. Since Schumpeter distinguished invention from innovation, the relationship between technology and economy has become an important topic of economic research.

When the theory of technological innovation was introduced into China, people initially thought that technological innovation was the first application of inventions in the commercial field. This understanding is reasonable in some content, but after careful analysis, two questions can be posed: first, should the results of an invention be commercialized without any changes or do they need to undergo certain changes during the process of commercialization; second, what does "first application" mean? Does it refer to the first application at international or national level? Or just at enterprise level? Economists regard technology as a "black box" which needs not to be opened, so they may not be interested in these questions, but for researchers of philosophy of technology and sociology, these are unavoidable questions. It was in this circumstance, after studying the technological innovation theory, the people who were originally against process theory began to think it was acceptable. After all, unless technology is regarded as a dynamic process, how could it innovate, and why would it need to? The application of technological invention to the commercial field is a process, the process of transformation of the technological invention and the process of turning technological invention into production technology and industrial technology. In this process, the purpose of the technology undergoes change, various elements of the technology undergo change, and the relationship between the technology and other technologies undergoes change, it is not the simple application of a fixed and unchanging invention to the commercial field.

Also, the first application of technological invention refers to the first time it is applied in commercial field. In being applied to commercial field for the first time, it will obtain intellectual property rights. If someone uses again a technological achievement that has been already applied for the first time, that person will infringe the intellectual property rights. Purchasing somebody else's intellectual property rights cannot be considered as innovation, it is called transfer of technology. Of course, on the basis of transfer of technology, innovation can continue.

From the dynamic perspective of technology, technological innovation includes the process of interaction between technology, system and management. It also includes the industrialization, commercialization and socialization of technology. In this process, technological invention is transformed into industrial technology, which is the perfected form of technology. Based on the above-mentioned diachronic analysis of technology from production to development, the transformation model is proposed as shown below.

Natural causality	+	Purposiveness	⟶	Technical principle
↓		↓		↓
Technical principle	+	Effectiveness	⟶	Technical invention
↓		↓		↓
Technical invention	+	Economic efficiency	⟶	Industrial technology

The diagram of 田-shape conversion model

In the transformation of these three processes, the transformation of purposiveness plays a leading role. The difference between scientific principles and technical principles lies in the purpose of human beings to break from natural constraints, resulting in the transformation of nature, to serve the purpose of human survival and development. But the purpose at this stage is still general. The transformation from the general purpose of technology to a specific one brings about the concretization of human purpose, only then we can speak of technological invention. Efficient technical invention has a variety of possibilities. Michael Faraday's law of electromagnetic induction produced the principle of the generator and the generator model. The generator is efficient in generating power. Thomas Edison applied it to illumination to create an electric lighting system. The electric power lighting system is a socio-economic technological system, thus, the effectiveness of the generator was put into play, and is used widely in society. In terms of electric lighting technology, Edison is both an inventor and an innovator. In fact, only when a technological invention has a more specific technological purpose, that is, to achieve a social and economic purpose, can it become industrial technology, thus achieving the purpose of exploitation, control and transformation of nature, and ultimately meet people's needs. Human beings proactive role is to constantly concretize the technological purpose. The aforementioned transformation from scientific principle to technical principle, and then to technical invention, as well as from technological invention to industrial technology, are the result of creative activities guided by human purpose.

4.3 Analysis of the Technological Value from the Process Theory of Technology

The research on technological innovation has promoted national technological innovation projects in China, resulting in further development of economic, management and philosophical research of technological innovation. Under these circumstances, some of the scholars engaged in the philosophy of science began to pay attention to the study of philosophy of technology. When philosophers of science take notice of technology, they attempted to model philosophy of technology upon the philosophy of science by analyzing technology as a kind of knowledge. On this model, epistemology was the core issue.

In the late 1990s, two professors, Huaxia Zhang and Zhilin Zhang, from Sun Yat-sen University in Guangzhou co-authored a paper called "The Research Program of Philosophy of Technology from the Perspective of Demarcation between Science and Technology." (*Studies on Dialectics of Nature*, 2001, 2), in which they pointed out that technology is a particular system of knowledge with its own characteristics, and that the core issue of philosophy of technology lies is epistemology. They stated that "the research on philosophy of technology should take a turn towards epistemology and logic of technology". In response, Professor Changshu Chen and myself co-authored the paper "Rethinking of the Research Program of Philosophy of Technology: Discussion with Professor Huaxia Zhang and Professor Zhilin Zhang" (*Studies on Dialectics of Nature*, 2001, 7). We proposed that technology is the combination of intelligent elements and substantial elements, as well as the activities and dynamic processes involved in combining these elements. The core content of philosophy of technology should not be epistemology, but axiology. Unfortunately, this significant debate was suspended due to a number of reasons. We researched the issue of whether technology is productive power, and how to transform it into productive power, all of which touches on the question of the value of technology, but the question of the value theory of technology has yet to become the subject of systematic reflection and subsequent publications. Since the completion of the book *On Industrial Technology* in 2005, the axiology of technology has become the main theme of my reflections.

If we focus on how technology can be transformed into productivity, of course we need to study the issue of the transformation of technological form, namely the industrialization of technology, because only through industrialization can technology be turned into practical productive force. However, the commercialization of technology should become the core content of the research on the question of technological value, because technology can only be changed from potential to reality after being commercialized. A discussion about the commercialization of technology includes two aspects: the commercialization of technology as commodity, and the commercialization of technological products. The former refers to the commercialization of technology itself, technical value, and the latter refers the commercialization of technology once it has been transformed into a product. The value of technology itself and the value of technological products are different but related. The value of technology itself includes the intrinsic value of technology and the value of the technological product, and from the point of view of the form of value and the process of transformation, they are different: first there is technological value, then there is the value of technology. The value of technology comes from the transformation of the technological value, and it realistically reflects the technological value. Before we made a distinction, when value hadn't been researched, both could be called "value"; but it is necessary to distinguish them, or else technological intrinsic value and the value of technological products could be confused.

Value is the relationship between the value subject and the value object. The value object (attribute, function, event, thing, behavior, etc.) possesses value if it can satisfy the needs of the value subject, or if it is meaningful, or if it meets the expectations of the subject. Technology as an object is valuable in as much it has specific functions and efficacy, and can satisfy human needs. From the point of view of industrialization and commercialization of technology, technology as a value object is constantly changing, it has different forms and each form should have different value; as a value subject, different forms of a technological subject does not mean one person or group, but multiple subjects, and different forms of one technology have different subjects. During the technological transformation process, being able to satisfy the needs of the subject of an earlier form of technology doesn't necessarily mean being able to satisfy the needs of the subject of a later form. Only if it can satisfy the needs of innovators and producers, and meet the needs of consumers as well, can we say that an innovation is successful. But things are far from over, consumers are also diverse and each has different value standards. An innovation that could satisfy the needs of some consumers will not necessarily be able to satisfy the needs of others. Britain's Daily Telegraph released a poll in January 8, 2009. 2500 adults took part in the poll. The results showed that the very popular Karaoke machine was chosen as the most annoying invention, the reason is that people singing Karaoke might be enjoying themselves, but the audience has to endure unpleasant noise. They even think of it as an antisocial invention, because 10 people singing Karaoke, means 150 people have to endure the torture. Inventor Rob Laura argues, "if you create a product that's so popular and so annoying, you'll be remembered as a great inventor." It can be seen that different value subjects will make different evaluation on the same value object, because the value preference is different. Because of the function of technology, it is possible to say that technology has instrumental value, value of means, intrinsic value, natural value and social value.

Hence the controversy between subjective value and objective value is raised. This controversy mainly stems from the fact that many people only analyze the intrinsic basis of technological value from the point of view of utility and efficacy. The so-called internal basis is efficacy, and the external value refers to personal value realized through scientific and technological activities, which is the real technological value. This equates the internal basis of technology to utility and effect. Logically speaking, there is nothing wrong in saying that technology is valuable because it has utility and efficacy. But the utility and efficacy of technology are a consequence of technology, not an inherent regulation of technological value. For example, a person who has a fever gets a penicillin injection and the fever is gone. We can say penicillin has value, and the fact that the fever was gone was surely caused by the utility of penicillin, but we cannot say that "no fever" is the intrinsic value of penicillin. The intrinsic value of penicillin is that it kills the bacteria or viruses which cause the fever. For penicillin, the power of sterilization and detoxification leads to the function and effect of bringing down a fever. Therefore, when analyzing the intrinsic regularity or intrinsic value of technological value, we can not only look at the utility, but also analyze the relationship between function and utility, so as to get a realistic conclusion.

From the analysis of utility and efficacy above we can see there are three elements that construct value.

First element is the purposiveness of technology. Purpose are human desire and demand, it's the wish to transform nature in order for people to satisfy their desires and needs. The goals setting up of human being is the pursuit of human value. The purpose is the primary factor to constitute the technical value. Therefore, technology has its value pursuit from the means and methods of realizing the purpose.

Second element is the natural functionality of conformity with the purpose. From a philosophical point of view, the natural law of conformity with the purpose is a basic principle of technology. Human beings remake nature according to the natural laws, and make natural things assume some functions to meet people's purpose. Because this kind of natural function satisfies people's purpose, so people can use it. The natural function conformity of the purpose is not only the social expression of human pursuit of value, but also the natural basis for the generation of technological value. What human beings use is the natural function, which is produced according to the natural law that satisfies the human' need.

Third element is the functional efficacy of artificial nature. If we say that the natural function has objective material qualities, or we view it as a natural attribute of technology, then the efficacy of the natural function stems from subjective human fondness, emotion and value preference, therefore it's subjective. As different people differ in their fondness, emotions and value preferences, the same natural function can be efficient in different ways. The principle of the point of minimal friction resistance allowed for the stones of the pyramids to be carried from the slope to the top, also can be used to make car tires, bearings, etc., and they all use the same natural function derived from the same principle. Pure natural function is also efficient and valuable to people, but it has no technological value, only natural value. Therefore, artificial nature can only achieve effectiveness if it satisfies the value needs of people.

From the perspective of process theory of technology, technological value is generated through the following process: "from purposiveness to natural function with purpose, then to effectiveness". All of the three elements are intrinsic elements or components of technical value and are the internal basis of technical value. If the internal basis must be analyzed through the concept of "intrinsic value", we cannot use concepts such as effectiveness, utility, or benefit to describe it, but should use the concept of efficiency, that is, the effectiveness of the natural function with purpose.

The intrinsic theory of technology regards "effectiveness" as the core of the intrinsic value of technology, neglecting the natural function determined by natural laws. There is no effectiveness without natural functionality. On the other hand, those who only focus on the natural attributes of technology seem to believe that the intrinsic value of technology is only the result of natural effects. What they neglect is the purposiveness of technology, as if purposiveness only existed under certain social conditions and while technology is in use. Both views are one-sided. These theorists acknowledge that technology has both natural and social attributes, but they do not take it into account in concrete analysis. The twofold nature is the inherent attribute of technology, so we must also recognize its existence in the analysis of technological value.

From the point of view of process theory of technology, in the process of transformation from generation to maturity, technology manifests itself in different forms. In specific space and time conditions, the value of certain forms of technology can be determined. But from the point of view of the process of transformation of the form, the value of different forms of technology is different, so the overall value of the technology cannot be determined prior to the completion of technology. We can't judge its value only by one of its forms. This requires us to carry out the value analysis of different forms of technology.

In the process of commercialization of technology, sample preparation is a key stage for technical products before entering the market. Trial production technology is a kind of experimental technology, so it can be regarded as a technological form. In this way, technological forms can be divided into five: technological creation, technological invention, sample production technology, production technology and industrial technology and the corresponding technological value form is shown in the following picture.

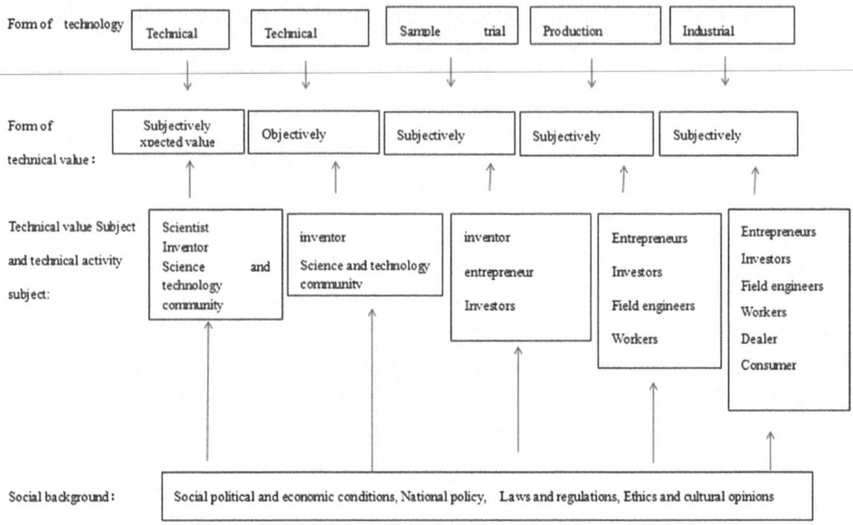

The symbiotic relationship between form of technology and form of technical value

With the transformation of technological form, the value form of technology also changes. The subject of technological activity is also the subject of technological value. When they create different forms of technology, they also impose different values on technology. There is a relationship of correspondence between technological form and technological value form, which can be called a symbiotic relationship between technology and value. This symbiosis is the result of the interaction between technology industrialization and technology commercialization. The mechanism of this interaction is mainly manifested in two aspects.

Firstly, the joint-formation mechanism of technological form and value form.

Technological instrumentalism seems to think that technology comes before application, that is, technological means emerge firstly, and then value is given to technology through application, because technology's value is neutral in itself. In fact, this is not true. Technology is endowed with value from the very beginning of its creation. Although it is subjective expected value, it is still a kind of value. This is especially true in the process of technological industrialization. As each stage of technology has different subjects of technological activity, technology gains a certain value at each stage. When technology changes from subjective to objective, from potential technology to reality technology, the value of technology also changes from subjective expectation to objective possibility or feasibility and from potential to real. According to the requirements of the market, the subject of technological industrialization activities needs to revise the goal constantly and seek new means, methods and ways to achieve that goal. In a way, it is like an endless process of establishing and solving equations. The subject of technological commercialization is constantly assessing the equation, adding variables to reflect the market demands, so that the equation has a particular solution. At each stage of the technical transformation an equation should be established, and then continuously evaluated and solved. If technology and its industrialization can be said to be the production of useful things by the subjects of technological activities through labor (in other words, they create value-in-use of technology), then the commercialization of technology creates things that can be exchanged and show their value in the exchange process. The value-in-use of technical goods reflects the relationship between man and things, and the value of exchange reflects interpersonal relationship. The interaction between industrialization and commercialization of technology creates the value and value-in-use, also creates different technological forms. This is the symbiotic mechanism of technological form and form of value.

Secondly, the joint-transformation mechanism of technological form and value form.

The transformation of the technical form must be accompanied by the transformation of the subjects of the technological activities. The latter have different value standards, so they give technology different value, and they transform technological value in the process of transformation of technological forms. This is the mechanism of joint-transformation of technical form and value of form. The value of technology includes economic value, ecological value and cultural value. During the process of transformation of technological form, some of the three values increase, the so-called "positive value", and some decrease, the so-called "negative value". In this way, the transformation of technological value is the process of increase or decrease of these three types of value. For inventors and engineers, the main concern is the functional value of technology, which includes the cultural value and the ecological value, but they must also pay attention to the economic value; as for entrepreneurs and investors, the main concern is the economic value. Technology has cultural value since it is first generated. As any technology needs to change the natural state, it also has ecological value. In the process of transformation of the technological form, technology are given value by different subjects, this value depends on the result of a "game" among value subjects of different activities. In every stage of the transformation of the technological form, there is a shift in the subjects of the activities, which give value according to their own sense of worth and assessment criteria, and also involves their own interests. In the transformation of technological form, what value the technology is endowed with depends on the result of the value game among various subjects, the value falls in the category of "futures", and no one can accurately evaluate the future value of technology. During the process of transformation of technology, the value game is not only a game of interests, it implies a compromise under certain specific social conditions, for every compromise, either the process of transformation of technology will reach its end, or it will take a wrong turn, maybe to the point of causing social harm. In the value game, the conflict may not only occur between the supply and demand sides of technology, but also among the economic, ecological and cultural values.

In order to solve these contradictions, the interest subjects must act according to national policies, laws and regulations, social ethical codes etc. From this point of view, A. Feenberg used interest analysis as his starting point and put forward the illuminating concept of "stakeholder". "A. Feenberg criticized the commonsense notion that technology equals efficiency and stated that efficiency alone cannot be turned into technology, it needs to be consistent with a specific interest. The people within a technological network all have their own interests, which in turn represent the interests of a social group" (Zhu, 2005). Indeed, the value game in the process of technological transformation is essentially a game of interests. The technological value of different forms is a compromise on the technological future value between various stakeholders. It is a temporary determination of it, within the uncertainty of technological value. Therefore, the value determined by this game is still uncertain, even hugely risky. The process of transformation of the technological form does not only imply the risk of whether technology itself can succeed or not, it's also an exploration of the risk of value uncertainty.

4.4 Waste Production: A Problem Worth Studying

For more than three decades, I have been thinking about the relationship between technology and production, discussing whether technology is a productive force, and how it can be transformed into productive force. Recently it occurred to me that besides technology and production, there is another element which should be taken into account: the waste issue. Considering the relation between technology and production from the perspective of waste issue, we can see there are many problems worth discussing.

The original meaning of *Shengchan* (生产, production) in Chinese language refers to child bearing and used to express the birth of a new life. But used in the economic context, it relates to people's livelihood rather than life. In the economic sense, the outcome of production is a pair of twins. One is the useful product, used for consumption in daily life; the other is useless waste, like wasted gas, water, slag muck or land, etc. Production is accompanied by the generation of waste. If there is no waste generated, there will be no production. Production is for consumption, and consumption is bound to produce waste, so no waste, no consumption. These are two basic empirical facts.

Human civilization began with the shift from hunting-gathering societies to the agricultural age. In the agricultural age there were two forms of production, one was agricultural production, which basically follows the laws of nature; the other was mining and hand-made, which are anti-nature forms of production. Development during the industrial age was carried out through this anti-nature extraction means. There's no waste in nature. Waste is produced in the process of agriculture and animal husbandry, but it can be recycled within the production process. On the contrary, in the case of mining, waste is left in the natural world, and nature itself cannot eliminate the waste, which leads to an environmental crisis, or rather, a life crisis, so it can be said that the life crisis is caused by the generation of waste.

People use technology to act on nature, and cause a variety of results or consequences according to the causality of nature. But people only focus on the purposive, useful and effective results. They keep the useful products and drop the useless ones. But the useless ones are also the outcomes of production. Waste is an artificial thing too. Although waste does not conform to human purpose, it still acts on mankind and nature according to the natural laws, and, in most cases, harms both people and nature. From the perspective of the inevitable bond between waste and production, there are two rules for the production of artificial nature: the one is the natural law that conforms to human purpose; the other is the natural law that does not conform to human purpose. The two play a role at the same time, and are originally in balance, but the neglect of the latter results in a deformed technological development, namely a highly developed production-consumption technology, but an extremely underdeveloped waste generation-waste removal technology. With the advent of mass production and mass consumption, we have found that the more production, the more consumption, and the more waste generated. As waste-removal technology is extremely backward, the accumulation of waste is becoming a catastrophe for humankind. It is essential for us to reflect on the fact that the lack of notions on waste-production and waste-removal leads to prejudice and inertia of thinking.

If we consider the natural law that conforms to human purpose as a basic principle of technology, then we could also say that the research on the natural law that does not conform to human purpose or it is transformation into one that does conform is also a basic principle of technology. The former promotes invention and development in the industry of production-consumption, the latter does the same in the industry of waste generation- management. According to the relationship among technology, production and waste, innovation in terms of production-consumption and innovation in terms of waste generation-management should keep pace with one another, but the actual situation is, the former is in leading position, the latter is carried out only to compensate the former out of sheer necessity.

Judging from the perspective of production-consumption, it seems that the update rate of technology and its production should follow the trend "the faster the better." The useful life of products in the machine age was one or two dozen years, in the electronic age it's only a few years. The life cycle of consumer goods changes shorter and shorter, there are more and more disposable consumer goods. Before a certain technological product has reached its maturity, a new one is launched, as is the case with the mobile phones, which undergo two generation updates per year. This way of thinking stems from consumer economy, because the faster the update in technology and its production, the larger the profit enterprises make, but that also means the generation of more waste. From the perspective of technology and waste generation, the correct choice is to extend the life cycle of products.

People seem to think that all high technology is good, but they seldom realize that the higher and more complex the technology, the greater the difficulties in the management of the waste, to the point that it's sometimes impossible. The waste generation rate of smelting a ton of iron is around 0.3–0.4 times the amount of iron, far below the waste-produce rate of smelting 1 gram of noble metals or rare elements, which is tens of thousands of times the amount. The quantity of waste and the waste generation-management ratio should be the assessment criterion for technological level and appropriateness, as well as a restricting factor to the development of technology.

In the past, Chinese philosophy of technology has emphasized how technology stimulates and contributes to the production of useful goods and services, that is, how it enhances productivity and economic development. But in China today we are increasingly aware that technology can also enhance waste and pollution. When the purpose of production is to consume, waste is naturally produced. In other words, not only should the positive value of technology be a theme of philosophy of technology, but the negative value should as well. It is the negative value that jeopardizes human life.

Many people lacked preoccupation about the quality of human production and thought that, as long as it improves quality of life and development, technology was a good thing. When it comes to human survival, production comes before livelihood, as there is no livelihood without production. Originally, nature already provided human beings with everything they needed to survive. After the separation of production and livelihood, the goal of production being, of course, livelihood, people didn't need to worry about air, water, clean soil, sunlight and such factors necessary to preserve life. But after the appearance of production methods, these basic necessities have been threatened by mass production and mass consumption. In the agricultural age, humankind already established the industrial clusters of production and consumption; during the industrial era, the industrial structure underwent a huge change and became an equipment manufacturing industry in which machines manufacture machines; the great change in the post-convention era lies in the appearance of the so-called environmental protection industry, its goal being the management of waste and the restoration of damaged ecological environments, in order to provide clean air, water and soil and improve human quality of life. Should we establish a new kind of industrial cluster that can restrict the existing industrial cluster of production-consumption and promote it at the same time? When this industrial cluster is established, it probably will be a huge technological combination of different levels of technology, illustrated as follows.

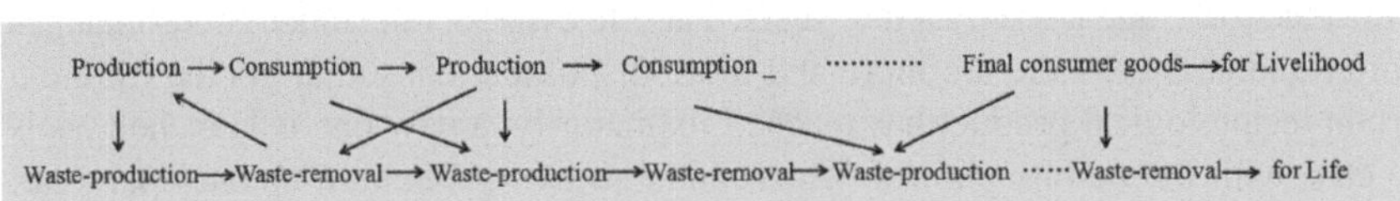

The industry of generation and removal of waste also uses technology as means, its products are also technological products, but they have two important features: one is its private nature, another one is its common nature. Enterprises recover misplaced resources through waste management, and make profit of it, which reflects its private nature. But most importantly, these enterprises must produce clear air, clean water and pollution-free soil, all of the three basic conditions for the preservation of life, this are common products. Can we let the market decide on the viability of such products? At this point, we have found a considerable amount of problems worth researching, including the notion of environmental protection (or maybe environmental generation would be a name more in accordance to reality).

How should we think philosophically about waste? Until now, this has not been a question asked in Chinese philosophy of technology. But this is a question that needs to become a major theme of philosophical reflection in relation to technology.

After considering philosophy of technology from the perspective of waste-production, we realized that we have only reflected on the side of technology and product-production but neglected the aspect of waste-production. How to keep the balance among production, waste, livelihood and life is a major issue in the development of technology and economy. That is the reason why I claimed that philosophy of technology is incomplete if it does not study the relationship between technology and waste. Similarly, economics is incomplete if it does not take the problem of waste into account.

The relationship between technology and waste is a problem that still holds many questions. Two years ago, when I talked with the American philosopher Carl Mitcham, I said that he was still a lion running in the academic forest while I was a wild crane in a solitary cloud, floating in the air and observing the problems on the ground. All I can do is to raise questions, but I cannot carry out the research myself. I hope that there will be more lions in the academic forest that will hear my call and keep on studying the problem of technology and waste-production.

References

Chen, C. (1994). On the transformation from science to technology—From the philosophical point of view. *Philosophical Researches, 11,* 35–40.

Gunter, S. B. (1993). *Heidegger analyzes the technology of the new era* (Z. Song, Trans.). Beijing: China Social Science Press.

Peng, F., Lin, J., & Li, Y. (2002). "World 3" and the essential characteristics of science and technology. *Philosophical Trends, 6,* 38–41.

Sun, Z. (1998). *General theory of philosophy* (p. 196). Shenyang: Liaoning People's Publishing House.

Xiao, F. (2005). The existence and non-existence of technology. *Science, Technology and Dialectics, 1,* 68–71.

Yuan, D. (2008). *Deyu Yuan's technology studies research collection*. Shenyang: Northeastern University Press.

Yuan, D., Ding, Y., & Ma, Q. (2005). *Industrial technology studies*. Shenyang: Northeastern University Press.

Zhu, C. (2005). *Criticism and construction: Research on Feenberg's critical theory of technology*. Ph.D. thesis.

Chapter 5
Ethics of Technology and Engineers' Professional Ethics

Zeyuan Liu and Guoyu Wang

Abstract There is a relatively popular view that regards ethics of technology as professional ethics and responsibility ethics of engineers. Although engineers must follow professional ethics relating to technical work and fulfill the corresponding social responsibilities of engineering technology, ethics of technology is not an equivalence of professional ethics or responsibility ethics. Nevertheless, from the genetic point of view, there is historical, realistic and theoretical kinship among ethics of technology, professional ethics and responsibility ethics of engineers. Ethics of technology has to be related to professional ethics and responsibility ethics of engineers. These three aspects are neither independent nor inclusive. Instead, they overlap or cross one another. Ethics of technology should be comprehended in terms of the following aspects: the relationship between man and nature, and inter-human relationship forming during technological activities; the impartible natural and social properties of technology; as well as the exploitation of man's intrinsic power using technology. Thus ethics of technology that fit modern technology can be found and rebuilt within this modern social construction of complex technological system. So, we should establish technological ethics based on the theories of "technology for harmony" and "technology for social construction".

Keywords Ethics of technology · Professional ethics · Harmony · Social construction

---

Z. Liu (✉)
Faculty of Humanities and Social Science, Dalian University of Technology, No. 2, Linggong Road, Ganjingzi District, Dalian, China
e-mail: liuzy@dlut.edu.cn

G. Wang
School of Philosophy, Fudan University, Yangpu District, No. 220, Handan Road, Shanghai, China
e-mail: wguoyu@fudan.edu.cn

Q. Wang (ed.), *Chinese Philosophy of Technology*, Philosophy of Engineering and Technology 34, https://doi.org/10.1007/978-981-15-1952-9_5

Within discussions of the ethics of technology, there is a relatively popular view that states the ethics of technology are the professional ethics and responsibility ethics of engineers. In 2003, for example, during the 1st Sino-German Symposium on the Ethics of Science and Technology, a Chinese representative pointed out that the ethics of technology research "should not neglect instances of rule violations or anomie in scientific and technological (S&T) activities." (Wang & Li, 2004) According to the pertinent study of an author in 2004, "as far as S&T morality is concerned, the ethics of technology is essentially a kind of professional ethics, which has general characteristics of professional ethics as well as its own special prescriptive nature." (Yang, 2004) In the following pages, we discuss three issues on the ethics of technology and professional ethics and present our viewpoints.

Firstly, we discuss what ethics of technology is not. We believe that the research object of the ethics of technology is not the professional ethics of engineers, and that the ethics of technology is not synonymous with professional ethics or the responsibility ethics of engineers. In the last three decades, in not only China but also Europe, the United States and other countries, data falsification, plagiarism and other violations of regulations and laws by not only unknown young scholars but also well-known academicians, academic authorities and Nobel Prize winners have been increasing and developed into a worldwide phenomenon. One example is Summerlin's fallacious claim about having transplanted the skin of a black mouse onto a white one at the Sloan Kettering Institute in New York in 1974, known thereafter as the "Summerlin Incident" or the "Watergate scandal of the scientific community." (Ren, Pang, & Yin, 2005) In 1980, Hongcheng Wang (王洪成) from Harbin, China falsely declared that he could make oil out of water. In 2002, Jan Hendrik Schon from the Bell Laboratory in the United States falsified data in research papers on molecular level nano-transistors. Known as the "Jan Hendrik Schon Incident," it is one of the biggest scandals in the scientific community in history. In 2005, South Korean embryonic stem cell scientist Woo-suk Hwang was discovered fabricating research data, which shocked international scientific circles. According to statistics, from January 1980 to January 1992, in *Nature* and *Science*, there were 266 papers that involved fallacious claims. (Fan, 1994) Given this background, it makes sense that some people have appealed to scholars to include scientific misbehavior, illegal activities and corruption within studies of the ethics of technology.

However, fabrication of data, falsification and plagiarism are neither unique phenomena of the modern high-tech era, nor are they phenomena limited to the scientific community. These are problems which have existed throughout history all over the world. Being practices which run counter to basic social and public morality, they are met with moral indignation and legal accusations and thus are despised. Sincerity and credibility comprise the basic moral character of a person. An old saying says that "sincerity is trust," and this forms the basis of decent human interactions. The thinkers in ancient China regarded sincerity as one of the five fundamental virtues, alongside kindness, righteousness, propriety, and wisdom. In English, the word "sincerity" originally meant "keeping the original form that has not been distorted, falsified or forged" (An, 2005), which also means honesty and truth. In German, the word "Wahrhaftigkeit" denotes sincerity and is one of the four pillars of Jürgen Habermas' discourse ethics.

In other words, sincerity as a moral norm in China and elsewhere, in modern and ancient times, is self-evident and a priori; its moral validity does not need the argumentation and justification of ethics and it is neither the responsibility of engineers in the technological age in particular nor the reflective object of the ethics of technology. So, should the ethics of technology be regarded as the responsibility ethics of engineers? German philosopher of technology and ethicist Günter Ropohl proceeded from the social system view of technology and examined the whole process from input to output of technology, from the birth of technological ideas to the use of technology and finally to the elimination of technological consequences. His complete and meticulous study points out that due to factors such as division of labor and cooperation, modern technological activities do not belong to sole individuals, so engineers as individuals cannot and should not take the full responsibility of technology. (Wang, Hubig, & Liu, 2007) German technology philosopher and ethicist Christoph Hubig analyzed the consequences of modern technology and pointed out that we should not discuss whether or not engineers should take social responsibility for technology but whether they can or cannot do so, for the following reasons:

(1) Any technological consequence shows probability, which means that there is no direct causal connection between the idea and purpose of technology and the technological effect.
(2) Because modern technological activities involve risks, when it comes to risks concerning the whole society or all of humanity, no individual can take full responsibility.
(3) Technological activities are systemic activities that fall under systems of social organization, so however outstanding an engineer is, he or she cannot be solely responsible for the technological activities of enterprises or organizations (Hubig, 2005). Under these circumstances, the interpretation of the ethics of technology as engineers' responsibilities can only be considered an empty doctrine.

There are many disastrous instances in history, such as the 2000 collapse of the Exhibition Center in Berlin, the US government's decision to use atomic bombs against Japan in the later period of the Second World War, NASA's *Challenger space shuttle* disaster, and the grave consequences of the Yellow River Sanmenxia reservoir in China, before all of which S&T experts had expressed their different views or warnings with a strong appeal to a sense of social responsibility. Yet these were all in vain, and this demonstrates the limitations of S&T experts as the subjects of liability for the ethics of technology.

Secondly, we will discuss what the ethics of technology is not and what it is. It is not equal to professional morality and ethics. It is, however, not only closely related to professional morality and ethics throughout history, but also has a realistic affinity for theory. Therefore, the ethics of technology must concern itself with the professional morality and ethics of engineering personnel.

The view of ethics of technology as the professional morality and ethics of scientists and engineers has social and historical roots. In view of the high social standing of scientists and engineers and the fact that misconduct happens now and then in S&T circles, the enhancement of construction for professional morality and education of ethics in the S&T field will have an extensive social impact and great practical significance, which is a major concern of the ethics of technology.

The developmental history of the ethics of technology in the US shows that there is kinship between the ethics of technology and engineers' professional ethics. Since the 1970s, the American Association for the Advancement of Science and the American Bar Association has jointly held the National Conference of Lawyers and Scientists during which moral norms in scientific research have been discussed. Afterwards, a series of reports on the social responsibility of S&T circles were published, such as *Scientific Freedom and Scientific Responsibility*, *Misconduct in Science and Engineering Research: Final Report* by the NSF and *Integrity in Scientific Research: Creating an Environment That Promotes Responsible Conduct* by the Committee on Assessing Integrity in Environment Research, the Institute of Medicine and others (Institute of Medicine and National Research Council, 2002). It can be seen that the development of the ethics of technology in the US since the 1970s has been closely related to an emphasis on the professional ethics of scientists and engineers.

In the US, it can be said that the ethics of technology originated from professional ethics. In his paper entitled "Achievements in Ethics of Technology", Professor Carl Mitcham reviewed the history of the ethics of technological research in the US and divided it into four phases. In the first phase, i.e., the latter half of 19th century, the community of occupational engineers formulated internal regulations. The second phase, i.e., the early 20th century, was a period of loyalty ethics, during which the ethical guidelines of AIEE, ASCE and ASME stressed business interests or corporate loyalty. The third phase began during the Second World War and emphasized that the main responsibilities of engineers are to ensure public safety, health and welfare. In 1932, the Engineers' Council for Professional Development was established, and it stressed that the main duty of engineers is to be responsible for public welfare, safety, and health. The fourth phase started in the 1970s and is a period of ethical education. Due to a series of incidents such as Ford's unsuccessful design of motors which caused deadly accidents, the Three Mile Island nuclear accident, and the movements of environmental protection and consumer protection, the ethics of technology was further stressed and became a course taught in classrooms.

The development of occupational communities and professional ethics in the US helped the research and development of the ethics of technology gain momentum in the past and at present, which shows that there is kinship between the ethics of technology and engineers' professional ethics. Theoretically then, the tendency to regard professional ethics as the ethics of technology is mainly because of the relationship between professional ethics and role ethics.

The idea of role responsibility may date back to ancient Greece, where people of different roles took different responsibilities. Role responsibility became the basis of virtue ethics and brought about the revival of virtue ethics, which stresses the morality of the actor. Christoph Hubig considers virtue ethics as the basis of the ethics of technology. On this basis, Verein Deutscher Ingenieure (VDI) formulated "Basic Principles of Engineering Ethics" as the engineers' code of conduct and professional duty in order to normalize and develop engineering and technology activities in line with human interests, which, to some extent, realized the unification of engineering ethics and the professional morality and responsibilities of engineers.

The ethics of technology, professional ethics and responsibility ethics are not equal to or independent of one another, nor is any of the three included in either one of the other two. There is historical and theoretical kinship among these three kinds of ethics. These three overlap or intersect with one another (Fig. 5.1).

Finally, we talk about what the ethics of technology and its research objects are. We believe that we should comprehend the essence of technology from the following aspects: the relationship between man and nature, interpersonal relationships formed during technological activities, and humanity's realization of its true power through the use of technology. From these aspects we shall find and reestablish an ethics of technology that is adaptable to this technological age and complex technological society.

We examine the research object of the ethics of technology from the following aspects: the essence of technology and the essence of humanity as reflected by it, as well as the essential relationship between technology and ethics. The research object of the ethics of technology is not technology or technological activities *per se*, but the relationship between man and nature, and interpersonal relationships formed during technological activities, as well as ethical issues caused by such activities. From the interaction of human-nature relationships and interpersonal

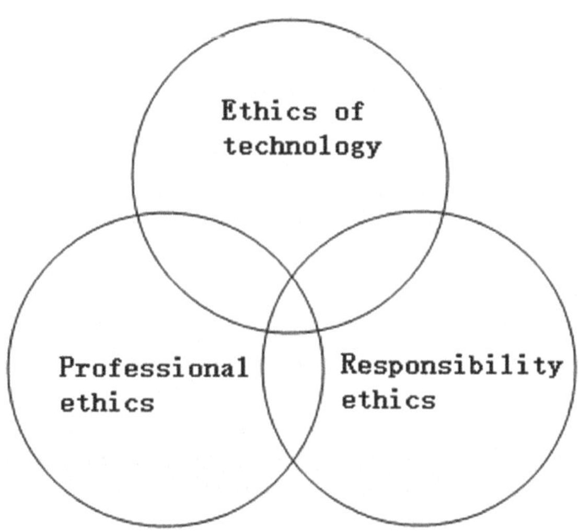

Fig. 5.1 The relationship among ethics of technology, professional ethics and responsibility ethics

relationships formed during technological activities, and at the same time in the two processes, i.e., the exteriorization of humans and the hominization of nature, the inseparable natural and social properties of technology come into being. Clearly, technology at its core reveals humanity's dynamic relationship with nature, and this kind of active relationship between humans and nature can not only be seen from activities remaking pyrophorus into artifacts but also from the rationality, ethics and freedom of humans in which lies the essence of man. Therefore, the essence of technology reflects and realizes the essence of humans. Thus, the essence of technology is connected with that of ethics, because man's rationality, morality and freedom are also the essential pursuits of ethics. These common pursuits and revelation of human essence are the basis of the claim of an intrinsic and essential relationship between technology and ethics.

However, in reality, the rapid development and strong power of modern technology exerts an increasingly widespread influence on humans, nature and society, and its consequences are both creative and destructive. (Zhao, 1997) There are unprecedented conflicts within the relationship between technology and ethics. Whether it is a challenge of modern technology against traditional ethics, or an ethical introspection of modern technology, all instances show the dual consequences of modern technology; ultimately, it is all due to technology dissimilation or technology paradoxes caused by modern society, and which is a reflection and result of human dissimilation. Therefore, fundamentally speaking, the appeal of modern ethics of technology is to proceed from the essence of technology, sublate human dissimilation in technological activities, let humans return to an essential state, and avoid technological dissimilation. Thus, in the process of advancing technology, a harmony is realized between technology and humans, technology and society, technology and nature, as well as between humans and nature, humans and humans, and humans with themselves. This is a kind of ethics of technology based on the teleology of "technology for harmony".

However, to achieve the aim of technological harmony, it is not enough to only consider technological processes; all technological subjects, processes, and objects should be taken into consideration, including aspects such as decision-making, creation, production, dissemination and usage of technology, and processes such as engineering processes, economic processes and cultural process so that social construction can be achieved; and it is not only engineers who should be responsible but all the social subjects of technology, including the decision-makers and organizers, creators and producers, demanders and users of technology as well as different sectors of society. Because technology can be regarded as social behavior and a way of living, the ethics of technology needs to research all social sectors involving technology, i.e., moral autonomy issues and ethics of responsibility issues of the social subjects of technology. This a kind of ethics of technology based on the theory of "social construction of technology".

As regards the social structural problems of technology, we can see from the knowledge system of engineering technology and its related disciplines that activities in engineering technology involve the following knowledge: the general technological theories and methods for engineering technology research and development; three levels of knowledge, i.e., natural science, technological science and engineering science; the interdisciplinary areas of social, technological, and scientific knowledge of natural science, technological science and social science such as innovation economics, technology sociology, ergonomics, technology ecology, technology management, technology aesthetics, and technology ethics (Liu & Cheng, 2009). It can also be seen from the status and role of ethics of technology in the complex cross-disciplinary technological knowledge system that without taking into consideration this system, technological knowledge for social constructions, and the relationship between technology and economy, society, ecology, aesthetics, or anthropology, it is impossible to comprehend the relationship between ethics and technology.

Similarly, if we regard technology as a complex system, it is not enough to only include the responsibilities of engineers and technicians in the studies of the ethics of technology, so we need to consider three levels of technological social subjects in such studies.

(1) The level of decision-makers and organizers, including politicians and entrepreneurs. They affect technology development and decision-making macroscopically in terms of politics and micro-economically respectively, thus they are the direct study objects of the ethics of technology. As technological activities intermingle with political activities, technological policies are political ones too. New energy development is related to national security and social stability, so it is not only a technical issue but also a political one. Transgenic technology is related to food and survival; nanotechnology has become the key in a new round of competition. Visionary politicians have to consider technological decision-making. Entrepreneurs are practitioners of technological development. From the selection of technical means, methods and resources to production through concrete technologies, entrepreneurs, as the decision-makers of technological production, are the direct study objects of ethics of technology.
(2) The level of creators and producers includes not only the conscience ethics of scientists and engineers as individuals but also the social responsibility of them as the keepers of this knowledge. It is here that the ethics of technology, occupational ethics, and responsibility ethics of engineers intersect with one another. The important thing is to find the intersection of ethics of technology, professional ethics and responsibility ethics. From the angle of engineers' professional ethics, the following should be studied separately: the relation between individuals' morality and values and professional ethics; the concrete manifestation of general social morals in the engineering and technology industry; particular ethical norms of various engineering and technology professions. As for the responsibility ethics of engineers, we should not only distinguish the scope and amount of responsibility, but also study how to assign technological responsibility, fault liability and legal responsibility.
(3) The level of demanders and users of technology includes all those who influence the developmental direction of technology and its consequences directly or

indirectly and should be the responsibility objects of the ethics of technology. Planes are a significant source of greenhouse gases, yet we are willing to choose or allow cheap flights, indirectly worsening the greenhouse effect. Holiday tourism is part of life in modern society, yet it is also a major cause of environmental pollution. We are keenly aware of a shortage of resources, but who is willing to stop using air conditioners in hot weather? All of these examples show that when technology has permeated every aspect of life and behavior, the objects of the ethics of technology are also the objects of general ethics and in this sense, the ethics of technology has the same aims and tasks as general ethics. Because of this, general ethical principles that regulate human behaviors also apply to the ethics of technology and can be the basic principles of the latter.

References

An, Y. (2005). "Sincerity" in Western culture and "Cheng" in Confucianism. *World Philosophy, 3,* 57–67.

Fan, H. (1994). Scientific falsification and its identification and prevention. *Journal of Dialectics of Nature, 16*(1), 25–33.

Hubig, C. (2005). Ethics of technology, in need of institutionalization. Translated by Guoyu Wang. *World Philosophy*, (4), 78–83.

Institute of Medicine and National Research Council. (2002). *Integrity in scientific research: Creating an environment that promotes responsible conduct.* Washington, DC: The National Academies Press. https://www.nap.edu/catalog/10430/integrity-in-scientific-research-creating-an-environment-that-promotes-responsible.

Liu, Z., & Cheng, G. (2009). On the innovative function and the strong country strategy of science of technology. *Research yearbook on science studies and management of science and technology in China 2007/2008.* Dalian: Dalian University of Technology Press.

Ren, B., Pang, Y., & Yin, C. (2005). *Illusions: 20 scientific seceptions that shocked the world.* Shanghai: Shanghai Culture Publishing House. http://scitech.people.com.cn/GB25509/4821139.html. Accessed 7 July 2007.

Wang, G., Hubig, C., & Liu, Z. (2007). Ethics of technology in the framework of social—technological system: On Ropohl's thoughts on utilitarianism. *Philosophical Researches, 6,* 78–85.

Wang, G., & Li, W. (2004). Cross-cultural dialogue on the ethics of science and technology: Commentary on German-Chinese symposium on ethics in science and technology. The original edition in *Social Sciences in China,* 1:102–108. The revised edition in: Li, W., Liu, Z., et al. (2006). *A study of German philosophy of technology.* Shenyang: Liaoning People's Publishing House.

Yang, H. (2004). What does ethics of science and technology study on. *Jianghan Tribune*, (2), 84–87.

Zhao, H. (1997). Sustainable development with Chinese characteristics: Green road research program. *Science of Science and Management of S& T, 18*(6), 5–8.

Chapter 6
The Triad of Science, Technology and Engineering

Bocong Li

Abstract Science, technology, and engineering are actually three different categories of social activities. According to the theoretical framework of the trichotomy of science, technology and engineering, scholars should pay attention not only to studying the characteristics, social functions and operational mechanisms of science, technology and engineering, but also to studying the connections, permeations, transformations, and interactions between them. From the perspective of the triad of science, technology and engineering, the philosophy of engineering should been seen as relative independent branch of philosophy different from the philosophy of science and philosophy of technology. The special issues are should been researched deeply in this area, in which the problem of "engineering levels" must be highly involved and intensively studied, that is mainly about "micro," "meso," and "macro" levels of engineering. Not only should scholars analyze the "numerous subject networks at the same level", but also should analyze and study the relations of "multi-level stereoscopic networks". Philosophy of engineering and "engineering studies" promote and permeate each other. Establishing "engineering studies" as an interdisciplinary and multidisciplinary field of research has important theoretical as well as practical significance.

Keywords Science · Technology · Engineering · Triad of science, technology, and engineering · Engineering studies

Since the New Culture Movement and the May 4th Movement, science and democracy have become resounding mottos and powerful trends in China. This "science" alongside "democracy", refers to the science in a broad sense, and its actual meaning and reference not only include science in a narrow sense, but also technology and engineering. (In this article, the term engineering refers to the engineering of natural objects in the usual sense, and the issue of "social engi-

B. Li (✉)
School of Humanities, University of Chinese Academy of Sciences, No. 19(A), Yuquan Road, Shijingshan District, Beijing 100049, China
e-mail: libocong@ucas.edu.cn

Q. Wang (ed.), *Chinese Philosophy of Technology*, Philosophy of Engineering and Technology 34, https://doi.org/10.1007/978-981-15-1952-9_6

neering" is not covered in this article). However, science also has a narrow definition, which means scientific researches. This "narrow science" cannot be confused with technology and engineering. This paper uses the term "science" to refer to this "narrow definition".

While science, technology, and engineering are actually three different categories of social activities, we also find that in daily life and within everyday language use, many people often confuse science with technology out of habit or lack of attention, or in some situations confuse technology with engineering. The view or concept which confuses science with technology and technology with engineering cannot be theoretically justified. In practice, such a view tends to have a misleading influence and adverse effects in terms of ideology, notions, and government policy.

It should be emphasized that the recognition of science, technology and engineering as three different categories of social activities in no way means denying the existence of the close connections between them. On the contrary, it is logical that only when the premise that they are three different categories of social activities is acknowledged can the connections and transformative relations between them arise and become prominent; for if science, technology and engineering are not three different objects but one, then, theoretically and logically, there will be no connections or transformative relations between them. That is, if we acknowledge this "three-in-one" view in theory, then the transformative process which leads from science to technology and from technology to engineering cannot exist.

The argument that regards science, technology and engineering as three different objects can be seen as a "triad" view of science, technology and engineering. According to the theoretical framework of the triad of science, technology and engineering, we should pay attention not only to studying the nature, characteristics, social functions and operational mechanisms of science, technology and engineering, but also to studying the connections, permeations, transformations, and interactions between them. It is self-evident that research, analysis and discussion of these issues are of important theoretical as well as practical significance.

For the convenience of analysis and argument, let us start with the mutual relation between science and technology.

6.1 On the "Monism" and "Dualism" Theories of Science and Technology

There are two different perspectives at home and abroad regarding the mutual relationship between science and technology: there is "monism" which states that science and technology as the same object, and there is "dualism" which takes science and technology as two distinct objects

In the monistic view concerning science and technology there are three subtypes, or to be more precise, there are three different and specific forms of expression.

The first form holds the "science-based" monistic view towards science and technology. The scholars who hold this point of view argue that "science and technology are all science." Although we should admit that many analyses and expositions from this view are reasonable and many insights are also worthy of attention, there is a fundamental flaw in this view: the serious underestimation or disregard for the independent status and features of technology. In this view, technology becomes oversimplified and is seen as the mere application of science, which in effect suggests that science encompasses technology and that technology is nothing but an appendage of science. We hold the view that while it is reasonable to emphasize the fundamental position and role of science, it is wrong to deny the independent status of technology. There is no doubt that technology has only "relative" rather than "absolute" independence, thus the word "relative" may itself be omitted; at the same time we wish to point out that while observing the relativity of independent status, we must acknowledge that all connections are conditional and therefore cannot be absolute.

The second form advocates a "technology-based" monistic view of science and technology. Scholars who hold this view argue that "science and technology are both technology." In this view, science is simplified to nothing but the "theoretical form" of technology. In effect, they advocate the inclusion of science into technology and view science as an appendage of technology. It is clear that this view also has a fundamental flaw, that is, it either seriously neglects or disregards the independent status and features of science. It is undoubtedly correct to emphasize the significance and uses of technology, but it is evidently wrong to deny the "independent status" of science because of this emphasis.

The third form holds the assertion that modern science and modern technology have combined and became one. For example, the theory of "scientific and technological revolution" (научно-техническая революция) advocated by some scholars in the former Soviet Union and the socialist countries in Eastern Europe agrees with this view. In the 1960s and 1970s, many scholars from the former Soviet Union and socialist countries in Eastern Europe published several treatises in this field. Scholars holding this view emphasize that although there were great differences between science and technology in the ancient world, science and technology have become inextricably integrated in modern society. Scholars from the former Soviet Union even created a new compound word in Russian (научно-технический- "science-technological") to show that "science" (наука) and "technology" (техника) are not two objects or two nouns but one unified new object or one "unified new word". They also assert that this "science and technology revolution" which integrates science and technology can only happen in socialist countries. Although this theory of "science and technology revolution" has become a relic of history, the monistic view of science and technology that asserts the merging of technology and science can still be seen from time to time. In our opinion, this view emphasizes that the relationship between science and technology in modern society is completely different from that in ancient society, which is absolutely correct. However, this by no means reasonably leads to the conclusion that modern science and modern technology have been unified. On the contrary, the two assertions that there is a close relationship between modern science and modern technology and that modern science and modern technology are still two separate objects are not mutually exclusive.

The American philosopher Joseph Agassi first published a view against the monistic view of science and technology and advocated a dualism of science and technology. In 1966, the Canadian philosopher of science Mario Bunge published an unusual article entitled "Technology as Applied Science" (Bunge, 1988). In this article, the term "philosophy of technology" appeared in English for the first time (the term had long since existed in German and Russian, but had not been used in English before), thus making the article an important document in the history of the philosophy of technology. However, the basic idea of this article was to strongly advocate technology as applied science and this view clearly carries with it a denial of the independence of technology. It should be emphasized that Bunge's view was not unique to him, but rather was a widespread and representative viewpoint. Bunge's view was immediately criticized by Joseph Agassi. Agassi published the article "Confusion about Science and Technology in General Science Philosophy" and criticized that the literature on philosophy of science … often confuses pure science and applied science and the differences between the both of them and technology. He argued that science and technology cannot be confused and criticized that all scientific philosophers of science treat science and technology identically (Agassi, 1988). The debate between Bunge and Agassi is actually the debate between "monism" and "dualism" in science and technology. How should we evaluate this debate and its subsequent social impact?

In my opinion, if we look at this debate from a purely theoretical point of view, it could be said that the dualistic view represented by Agassi clearly prevails and thus constitutes the mainstream view in the field of philosophy of science and technology. The telltale sign of this is the "academic society" of philosophy of technology that was soon established in the United States. In addition, there were many dissertations and treatises, as well as some professional magazines that were subsequently published in the field of philosophy of technology. However, if we view this problem from different angles, particularly that of the general opinion, we see that the situation is different, since among the communities or even "scholar groups" from other fields, the phenomenon of confusing pure science with technology and ignoring the differences between them are not rare, but rather common and even more influential and more widespread. The manifestations of these two aspects seem contradictory, but it is actually the real situation within social life.

The two aspects mentioned above exist not only in developed countries, but also in China.

In China, Professor Changshu Chen was the first to clearly insist that we not confuse science with technology and firmly advocated the study of the differentiation between science and technology. In 1982, Chen published an article entitled "Unity and Differences in Science and Technology" in *Guangming Daily*, pointing out that science and technology are "two categories" "two values" "two revolutions" "two innovations" and "two tiers of management" (Chen, 2002a, 9–14). This article is recognized as a pioneering work as well as the genesis of the philosophy of technology in China. While in the process of writing the article, Professor Chen explicitly pointed out that if we neglect or underestimate the difference between science and technology, "it will not only hinder an understanding of the relationship between science and technology, but also obstruct the formulation and implementation of correct policies of science and technology." The analysis and understanding of this problem by Professor Chen significantly exceeds that of Agassi, although Chen started the study on the difference between science and technology only much later. It should also be pointed out that Chen arrived at these views and conclusions from his own approach.

Chen's view has become mainstream and dominant in Chinese philosophy of technology. However, since the circle of philosophy of technology is limited, if we view this problem from the perspective of general public opinion or the general academic community outside this circle, we need also to admit that the viewpoint put forward by Chen has not become the dominant belief, and that the phenomenon of confusing science with technology has always been common and widespread.

The above are the theoretical forms of the monistic and dualistic views of science and technology. What should be emphasized here is that for many who are not engaged in theoretical work or theoretical research, they often fail to clearly or consciously think about the theoretical issues of science and technology monism or dualism, yet when they use everyday language, they often confuse science with technology and unknowingly adopt the monistic notion of science and technology. A typical and trite reflection of this situation is found in the use of the compound word in Chinese for "science and technology" (abbreviated to two Chinese characters "科技" instead of what would ordinarily be four characters). Mr. Dayou Wu (吴大猷) said that in Taiwan, "ordinary folks often lump science together with technology" (Wu, 1996). In the mainland, when using the term "science and technology" or the abbreviation of "S&T", many people also often fail to realize the need to make a distinction between the two. At this point, it is worth mentioning that one can also find in Japan the phenomenon that "science and technology are treated as homogeneous, and they are used together as terminology on various occasions" (Li, 2002a, 2002b).

It is exactly because the reality that the monistic point of view of science and technology still has a great influence among the general public so far that some scholars in China have felt and continue to feel obliged to write to emphasize and analyze the differences between science and technology in the late 20th century and early 21st century. There is still a need to assert that science and technology cannot be confused with one another because if "technology and science are confused, there will be a negative impact on both" (Bocong Li et al.).

6.2 The Differentiation of and Connections between Technology and Engineering

It should be noted that in addition to questions of whether or not there is an essential distinction between science and technology, there is another issue as to how to identify the mutual relationship between technology and engineering. On this issue, and similar to the relationship between science and technology, there are also many situations where technology and engineering are confused.

The tendency to confuse technology with engineering is common to both China and foreign countries. For example, Professor Hans Poser from Germany considered technology and engineering to be equivalent (Poser, 1998), as evidenced when he said: "engineering and technology, engineering science or technology science, or the philosophy of technology or the philosophy of engineering need not necessarily been differentiated from one other and are furthermore difficult to distinguish. Engineering philosophy is equivalent to the philosophy of technology, and it is not necessary to differentiate so specifically between engineering and technology" (Poser, Liu, & Li, 2003). Professor Xukun Wang (王续琨), a Chinese scholar, agreed with Poser's view and in his dialogue with Poser he said, "In China, I understand it that engineering is the process by which we reform the world and technology is the methods of doing so. Both of these are two aspects of the same issue. Engineering and technology are inseparable and without technology, engineering could not be conducted. Without engineering, technology would have no support." Professor Zhaoyou Li (李兆友) and Professor Zeyuan Liu have also agreed with Professor Poser's view and argue that "philosophy of engineering is equivalent to the philosophy of technology", and that "it is not necessary and indeed difficult to distinguish between philosophy of engineering and philosophy of technology." They disagree with the trichotomy of science, technology and engineering and state that "technology and engineering cannot be understood as parallel concepts" (Li & Liu, 2003).

Professor Changshu Chen discussed the relationship between technology and engineering in his book *Introduction to the Philosophy of Technology*, but the book did not make this issue very clear. In 2002, Professor Chen published a paper entitled "Attach Importance to Engineering, Engineering Technology and Engineers", wherein he clearly affirms that engineering is the object which should not be confused with technology. After enumerating ten features of engineering, Chen concludes, "engineering is not only closely related to but also considerably different from technology. Engineering has its own relative independence and particularity and as such, engineering problems require specialized discussion" (Chen, 2002b, 30).

In recent years, I have further studied the relationship between engineering and technology. On this issue, I think it is worthwhile to stress in particular the following viewpoints.

Firstly, there is no engineering without technology and thus there is a close relationship between engineering and technology.

Secondly, there is no such thing as "pure technology" in engineering and therefore technology and engineering must never be confused. There exists within engineering activity not only the element of technology, (please note that the technology in engineering activities is not technology in a "general form", nor is it in a "laboratory form", but rather a form better named as "engineering technology"), but also other elements such as management factors, economic factors, institutional factors, social factors (including politics and law), ethical factors, etc., which means that engineering is the integration of technological factor and non-technical factors. It is because engineering activities are by no means "purely technology" activities that when organizing and carrying out engineering activities, not only the chief engineer but also the general director (or general manager), chief designer, chief accountant, technicians and workers are all needed. Technical elements and components are undoubtedly important in engineering activities, but the importance of other elements, especially economic and managerial elements, are frequently not seen as great as the importance of technical aspects. Among the ten features of engineering activities that Professor Chen described, most of them are characterized by "non-technical factors" or "non-technical elements". It should be emphasized that in engineering activities, "non-technical factors" in a project are by no means secondary or non-essential, rather in many cases "non-technical" elements (such as economic factors or political factors) often turn out to be essential and decisive. Therefore, the view that regards engineering and technology as "inseparable" and denies the independence of engineering is unsuitable.

Thirdly, technology can be "applied" to engineering. The process of the "application" is a process of transformation. It is precisely because the "application" process is a process of transformation, and that after this transformation process takes place there must be new substance, that to admit that engineering is an "application" of technology should not be a reason to confuse technology with engineering, but on the contrary be a reason to affirm that there are essential differences between technology and engineering. It should be noted that this process of transformation is complex and conditional, and so ideas and approaches which simplify this process of transformation and transformation relationships are in error.

Fourthly, an engineering project needs to choose and integrate technology. At the planning and design stages of engineering activities, the executive agents need to select and integrate existing technologies according to the objectives of the engineering activity. From this relationship of choosing and integrating, we can also clearly see that technology and engineering are by no means the same thing or the same object. In the 1970s, E. F. Schumacher wrote the famous book *Small is Beautiful*. While this article will not deal with Schumacher's specific views, what I wish to say is that from the point of view of philosophy of engineering, one of the important points of significance in this book is that it highlights the view that technology is a "selected" and "integrated" object. That is to say, specific technology must be selected according to an engineering need and integrated into industrial and engineering activities according to the social environment and conditions of the situation. Thus, he criticizes and denies the approach, standards, and beliefs of "pure technology". The relationship among application, selection, and integration above reflect not only the connections between technology and engineering, but also the difference between technology and engineering (here, I would like to repeat and stress that "interrelatedness", "inter-permeation" and "interaction" are all based and predicated upon "distinction").

Fifthly, it is precisely because engineering and technology are essentially distinct social activities that technical and engineering activities require different evaluation criteria (probably nobody would agree that technical activities and engineering activities can adopt the same evaluation criteria). If instead we must admit that technology and engineering have different evaluation criteria, it will be inevitable for us to then deduce that technology and engineering are two different categories of social activities.

Above are stated the relationships between science and technology and between technology and engineering. The following content will focus on the three objects of science, technology, and engineering together for comparative analysis.

6.3 Viewpoints of Triad of Science, Technology and Engineering

Professor Chen criticized the tendency to confuse science, technology, and engineering. He said: "People often confuse engineering with technology, which is similar to the confusion of science with technology. Engineering and technology are often treated as the same object, and it seems that the two are identical to many people. From the beginning science and technology have differed in principle, and there are also considerable differences between engineering and technology. However, the differences between science, technology and engineering are not readily acknowledged" (Chen, 2002b, 28–29).

For Professor Chen, it was not easy to persuade people to agree that there are essential differences between science, technology and engineering. Therefore, it is necessary for us to think deeply about and research the issues in this field, and to devote ourselves to elaborating on the "triad" of science, technology and engineering.

The so-called "triad" of science, technology and engineering is based on the recognition of and affirmation for the distinctive characteristics of science, technology and engineering, seen as three different objects and social activities, yet also being closely related to each other at the same time. The different natures and characteristics of science, technology and engineering can be illustrated through analysis and comparison of the following aspects.

(1) From the perspective of content and nature, scientific activities, technological activities and engineering activities are three different kinds of social activities. The core of scientific research activities is discoveries, the core of technological activities is inventions, and the core of engineering activities is construction.
(2) As regards the nature and type of "achievements", scientific "achievements", technological "achievements" and engineering "achievements" are of different characteristics and sorts: the main forms of scientific achievements are scientific concepts, scientific laws, and scientific theories, and thesis and writings, which are treasures for all mankind and are "public knowledge" rather than knowledge that any person may have "privately"; the main forms of technological

achievements are inventions, patents, technical know-how, patent literature, drawings, formulas (and, of course, technical literature and papers as well), which tend to be "private knowledge" for a specific amount of time as they are associated with "patents" (of course, we also acknowledge that technical "achievements" may also be a sample or prototype of a material form); the main form of engineering achievements is artifacts or a physical facility, which, in general, is a directly material wealth itself. Besides instances of public engineering, the "goods" produced as the achievement of engineering are not "public" but "belong to" a specific "subject".

(3) As regards "activity subjects", "social roles" and "community structure", the main subjects and agents of scientific, technological and engineering activities are different. The "main agents of scientific activities" (or "role" as used in sociology) are scientists, the "main agents of technological activities" are inventors, and the "main agents of engineering activities" are entrepreneurs, engineers, investors, managers and workers. Within philosophy of science, Kuhn is known for his work on the "scientific community". In the philosophy of science, the issue of the scientific community has already become an important research topic. By contrast, the study of the "technological community" and "engineering community" is still in its infancy. Clearly, studying the "technological community" and "engineering community" is by no means less significant than studying the "scientific community". Professor Chen suggests that we should not only pay attention to the study of scientists and inventors, but also pay attention to the study of "engineers" (Chen, 2002b, 33). I fully agree with his argument.

(4) The characteristics of the subjects of science, technology, and engineering as well as the thinking patterns behind them are also not to be confused. The object of science is scientific "laws", characterized by universality and repeatability. The object of technology is universal and repeatable "methods". Any scientific law or technological method must have a certain degree of "repeatability" and must be "universally valid". There is no such thing as "one-time" scientific laws or technical methods. In engineering activities, however, the situation is completely different since any project (please note that project here refers to "engineering projects" instead of "engineering science" and "engineering technologies") is by necessity a unique and individual project. As Professor Chen points out, every project is a "singular" object of a "one-time" nature, such as the Qinghai-Tibet Railway Project and the Nanjing Yangtze River Bridge Construction Project (Chen, 2002b, 29). Engineering thinking, technological thinking and scientific thinking are three kinds of thinking patterns.

(5) As regards the aspect of institution, scientific, technological and engineering institutions are three different kinds of social "institutions". They have different arrangements, environments, operation modes and activity standards. They have different evaluation criteria and evolutionary paths, different management principles, different development modes as well as different goals and directions.

(6) Since science, technology and engineering are three different kinds of social activities that have different positions and uses in society, the state and the government need to formulate policies for science, technology and engineering research that differ in content and function "separately". All among these three policies are indispensable and should not be replaced.
(7) From the perspective of culture studies and communication studies, science culture, technology culture and engineering culture also have different connotations and characteristics. The "public understanding of science", "public understanding of technology", and "public understanding of engineering" are also "irreplaceable", and each has its own special content, meaning and social function.
(8) Emphasizing that there are essential differences between science, technology and engineering by no means denies the close connections between them. On the contrary, it is precisely because each has its own unique nature and its own special uses which cannot be replaced by other activities that their problems of "position", "status" and "connection", as well as problems of the transition from science to technology and from technology to engineering are all brought forward from theory, practice and policy.

At this point, I would like to give some explanations of three issues. One is about the meaning of the term "engineering technology": if the meaning of "science technology" is often understood as "science and technology", then it should be noted that the term "engineering technology" does not generally refer to "engineering and technology", but to "technology in engineering forms" or "technology used in engineering". The second issue concerns the relationship between engineering and economy. According to K. Marx, engineering activities concern concrete labour and use value, while economic activities concern abstract labour and value.

Many engineering projects are often divided into many "sub-projects". A project may be undertaken by one enterprise alone or by several enterprises. Indeed, there are also some cases in which one enterprise undertakes several projects at the same time. The third issue concerns the relationship between engineering and industry. "Industry" is a general term for many engineering activities of similar types.

6.4 Three Levels of Engineering: Micro, Meso, and Macro

While researching engineering, we must be highly concerned with research on the problem of "engineering levels". Although this problem is "pervasive" in the real world, it remains "vague and intangible" in theoretical studies and occupies an "ignored" corner. This condition must be changed. In the field of philosophy of engineering and the theory of the evolution of engineering, the problem of "engineering levels" must be highly involved and intensively studied.

The so-called problem of "engineering levels" is mainly about "micro," "meso," and "macro" levels of engineering. It can also be called a problem of engineering scope or scale, as well as a problem of "viewpoint" or "research framework" from the perspective of theoretical studies or observers. This article will do some preliminary analysis of engineering levels and the "micro-meso-macro" research framework mainly from the angle of philosophy of engineering and the theory of the evolution of engineering.

1. How to Divide the Three Levels of Engineering

The distinction of "micro" and "macro" originates from physics. While in general, many people often distinguish between "micro" and "macro" in the sense of relative size. For example, what can only be seen under the microscope as a "microscopic object". However, some scholars distinguish between "micro" and "macro" in a more special sense, calling "fundamental particle" world "micro world". In the "micro world", the laws of quantum mechanics play a role; while in the "macro world", Newtonian physics plays a role.

Later, economists borrowed the two terms "micro" and "macro" from physics. However, what economists borrowed is just the literal meaning without "introducing" the scale standards of "micro" and "macro" in physics. More specifically, the "micro" in economics and the "micro" in physics are based on completely different standards, and their specific meanings are also completely different – the "microscopic objects" in economics are all "macro objects" in physics. Peigang Zhang (张培刚), a famous economist in China, said, "The 'macro' of macroeconomics, the 'micro' of microeconomics, was originally a concept used in natural science, especially in physics. It was meant to be 'grand' and 'small'. It was transferred to economics, first in the early 1930s in this century (referring to the 20th century), but it was limited to individual occasions. After the end of the Second World War, especially from 60s to 70s, the study of 'macro' and 'micro' became prevalent" (Zhang, 1997).

There is an encyclopedia of economics that says, the term "micro" means small, and microeconomics means economics in a small range. The optimization behavior of individual units such as family and business is the foundation of microeconomics. In correspondence with "micro", the word "macro" refers to "extensive", and macroeconomics refers to large-scale economics. Macroeconomists are concerned with such issues as aggregate production, total employment and total unemployment, total level and speed of price changes, as well as economic growth rate (Greenwald, 1992). In the field of economics and economic analysis, since individuals and families are the "micro-subjects" of consumption activities and enterprises are the "micro-subjects" of production activities. Thus, in the economics, the microeconomic study objects are defined as research on the economic activities of individuals, families, and businesses; and the so-called macroeconomics is defined as the study of economic activities on the "national scale" or even the "world scale". Later, economists proposed the concept of "meso" to refer to the economic phenomena and economic theory of industries, industries or "regions" between "micro" and "macro".

In business ethics study[1] field, some scholars have also paid attention to this so-called "micro", "meso," and "macro". It is worth noting that the division and definition of the so-called "micro", "meso" and "macro" levels by ethicists are quite different from those of economists. For example, Georges Enderle said, "In order to identify the subjects of responsibility as concretely as possible, three different levels of acting are proposed, each of which includes actors with their respective objectives, interests and motivations: the micro-level, meso-level and macro-level. At the micro-level the focus is on the individual, that is, what actors, as employee or employer, colleague or manager, consumer, supplier or investor, do, can do and ought to do in order to perceive and assume their ethical responsibility. At this level we also consider groups, composed of small numbers of individuals and without organizational structures, making collective decisions and taking collective actions. At the meso-level we consider the decision-making and action of economic organizations, chiefly business firms, but also trade unions, consumer organizations and professional associations. Finally, at the macro-level we consider the economic system as such and the shaping of the overall economic conditions of business: the economic order with its multiple institutions and economic, financial and social policies" (Enderle, 2002). Between the definitions of Enderle and economists on "micro", "meso" and "macro", many distinctions can be found, among which three aspects are the most important: the first is that Enderle only recognizes the individual as the micro ethical subject; the second is that "manufacturer" in economics that is classified in the microeconomics is defined as "meso" level objects; the third is that Enderle's explanation to macro-level is quite different from that of economics. The reason for these differences is not because economic ethicists do not understand the relevant concepts of economics; on the contrary, economic ethicists are clearly aware of the differences. For example, in his work, Enderle explicitly acknowledges that "the three-level definition" adopted by economic ethics is quite different from the "definitions" of micro and macro in economics. He also indicates that "the central point of this three-level conception is to perceive the links between decision-making, acting and responsibility as concretely as possible and to provide conceptual room for addressing the differences and conflicts of objectives, interests and motivations which are located at the interfaces between different levels" (Enderle, 2002). It can be seen that because the discipline nature and subject focus of economics and ethics are different, there are different views on dividing the three levels—"micro", "meso" and "macro". Defining the scale, scope, and specific content of the "levels" of micro, meso, and macro, this article mainly adopts the relevant definition of economists, that is to say, defines "individual" and "enterprise" as the "micro-level" subject of engineering activities, "industries", "regions" and "industrial clusters" as the "meso-level", and the "state" and "global" scope as the "macro-level".

[1]Business-ethics, is translated into "经济伦理学" by Chinese scholars. What should be noted in this translation is that the corresponding English word of "经济" in "经济伦理学" is not "economy" but "business".

2. The "Micro Subject" of Engineering Activities and Its Evolution

Since any engineering activity is a "collective" activity and in modern society enterprises are a common subject form of engineering activities; [2]therefore, when analyzing and discussing the micro-subjects of engineering activities, this session "implements" this issue into the analysis and discussion of "individual" and "enterprise" and its evolution in engineering activities.

(1) The role division, role function, role structure of individuals, and their evolution in engineering activities

Engineering activities are collective activities of division of labor and cooperation.[3] When talking about division of labor, many people cannot help thinking of Adam Smith's description and comment on the division of labor at manual factories in *The Wealth of Nations*. As an economist, Adam Smith analyzes the effect and meaning of division of labor mainly from an economic perspective. But from the perspective of philosophy of engineering, the importance of division of labor is first expressed as the premise and foundation of carrying out engineering activities. If there is no division of labor, it is impossible to carry out engineering activities. Karl Marx said, "How far the productive forces of a nation are developed is shown most manifestly by the extent to which the division of labor has been carried. Each new productive force, insofar as it is not merely a quantitative extension of productive forces already known (for instance the bringing into cultivation of fresh land), causes a further development of the division of labor" (Marx & Engels, 1998). The basic feature of engineering is that many individuals in the engineering community conduct engineering activities in both divisional and cooperative manner, which are inseparable with each other. The importance of the cooperation with a division of labor can be analyzed from two aspects. From the one side, many individuals participating in engineering activities must be in coordination and cooperation with a division of labor to ensure certain engineering activities; from the other side, engineering activities cannot be implemented if lake of necessary division of labor and corresponding cooperation. The division of labor is the inherent necessary, prerequisite and basic condition of engineering activities, and inevitably becomes the typical manifestation and basic content of engineering evolution. Therefore, the evolutional history of division of labor becomes one of the main contents of engineering evolutional history.

The following facts below must be of great concern when examining the evolution of human individual evolution.

---

[2]In modern society, the specific forms of the main body of engineering activities are diverse. This article cannot analyze and elaborate on it more specifically, but can only be represented by enterprises.

[3]There are two major types of division of labor: intra-organizational division of labor and social division of labor. When conducting micro-analysis, it will only deal with issues related to "intra-organizational division of labor", and when it comes to meso-analysis and macro-analysis, it will involve issues related to "social division of labor".

In history, since apes evolved into human beings, human beings have experienced a long evolutionary course lasting for at least two to three million years, in which the physical structures and functions of human body (such as the cranial capacity size, the details of the specific anatomy of the human hand) has changed a lot. However, if the scope of the study is limited to a historical period of about 10,000 years—especially for thousands of years of written records, an obvious fact can be seen that during this period, the physical structures and functions of human body change little, or even to say nearly change nothing. However, this short period of time is "the most sharp and deep" time in humans' evolutionary course.

The key point is that during this period, though the individual person's "physique and physiological characteristics" have no remarkable changes, the "condition of labor division", "role capacity" and "role structure" of individuals had changed tremendously.

History vividly tells us that the history of division of labor is the one of most important contents in the whole engineering evolutionary history. A Chinese ancient classic named *Kao Gong Ji* (考工记) compiled more than two thousand years ago records particularly over thirty crafts in the official handicraft industry of the Zhou Dynasty, "there are seven kinds of carpenters, six kinds of metal fabricators, five kinds of tanners, five kinds of strainers, five kinds of scraped finishers, two kinds of potters" (Wen, 1988). A. Smith, who lived more than 200 years ago, used three chapters to make a specific, deep and detailed analysis and study on division of labor in his masterpiece *The Wealth of Nations*. His narrative of "division of labor" in fibula manufacturing at that time has been cited frequently by later generations as a typical example of the situation and effect of the division of labor in handcraft industry then. A. Smith said, "The greatest improvements in the productive powers of labor, and the greater part of the skill, dexterity, and judgment, with which it is anywhere directed, or applied, seem to have been the effects of the division of labor" (Smith, 2003, 9). Karl Marx also talked in detail the situation of the division of labor in the watch industry in *Capital*, "Formerly the individual creation of a craftsman from Nuremberg, the watch has been transformed into the social product of an immense number of specialized workers, such as mainspring makers, dial makers, spiral-spring makers, jewelled hole makers, ruby lever makers, hand makers, case makers, screw makers, gilders. Then there are numerous subdivisions, such as wheel makers (with a further division between brass and steel), pin makers, … (the citer's note: 26 craft subcategories listed in total)" (Marx, 1992). It should be emphasized that these examples can not only be the typical cases for economists to study economic problems, but also for studying technical divisions of labor in engineering activities. It is not difficult to draw a conclusion from numerous similar cases that in the developmental process of engineering activities and manufacturing industries, the change, differentiation, development and evolution of "division of labor" are one of the most important and prominent manifestations that reflect and show how individual effects and functions evolve in engineering activities.

How to understand the reasons, nature and role of division of labor on earth are not only an economic issue, but also are issues of engineering, philosophy, sociology and politics. Under the condition of division of labor, numerous individuals become different "members" and "roles" in "community" due to the results of "division of labor" and the distinction of "positions". The engineering community is composed of different "positions" or "roles" with different duties. Only when a person has the knowledge and ability corresponding to the "requirements" of the "positions", can the person take on the relevant "role".

When recognizing and analyzing the problem of "humanity", many theorists stress on taking different human individuals as "homogeneous" and "indifferent" individuals from the perspective of humane studies, however, when joining up to conduct engineering activities, "these numerous homogeneous individuals" are inevitably transformed into "heterogeneous" with different functions through "division of labor". Then, is the difference in positions or role due to some differences in the essential characteristics of the person? The answer is no. A. Smith said, "the difference of natural talents in different men is, in reality, much less than we are aware of; and the very different genius which appears to distinguish men of different professions, when grown up to maturity, is not upon many occasions so much the cause, as the effect of the division of labor" (Smith, 2003, 25). A. Smith and many scholars have clearly pointed out that as a result of the division of labor, those who specialize in a certain working field of division of labor can develop his or her "related capabilities" unprecedentedly.

When observing the overall evolutionary trend or the overall evolutionary features of the evolutionary course of division of labor in human engineering activities, it is most noteworthy that in the period of handcraft industry and "Fordism", the division of labor was developing as a whole along the direction of "more and more specific", yet when entering into the so-called "post-Fordism" period, the development of division of labor diverted to "generalists" who need to and could undertake multiple positions, that is, the trend of requiring individuals to "develop in an all-round way" in some sense appears.

From the perspectives of philosophy, history and sociology, not only the relationship between individuals and groups, but also its closely-connected relation between individual "division of labor and development on position ability" and "personal comprehensive development" is an important and complex issue. It's reasonable to expect that in the future evolutionary process the individual "comprehensive development" will surely get more attention and new progress while the individual "division of labor" continues to develop.

(2) The "Organization Forms" of "Micro-Production Subjects" and their evolutions in engineering activities

In a long period since human's history entered into agricultural society and handcraft industry society be differentiation, humans do "manufacturing activities" besides agricultural activities mainly in the organization form of handcraft workshop.[4] In the social environment of that period, the handcraft workshop became the chief organization form of "microscopic subject of manufacturing activities", which, however, changed deeply in the development of capitalism. First, "handcraft factories" based on simple cooperation and cooperation of division of labor emerged. "Handcraft factories based on cooperation of division of labor develop gradually through two forms: one is to disintegrate and simplify these handcraft industries, through uniting the handicraftsmen of different kinds into the same factory, until they become complementary partial operations in the same commodity production; the other is to gather many handicraftsmen of the same kind into the same factory to gradually divide the homogeneous handcraft industry into various operations and make every operation independent enough to be specialized skills of partial laborers." "As the typical form, handcraft factories based on the division of labor focused mainly on the industries requiring cooperation at the beginning, such as textile, mining, metallurgy, and shipbuilding" (Xie, 2005). Afterwards, more industries are involved.

From the perspective of theoretical analysis and historical evolution, the most revolutionary event of industrial production activities in the evolutionary course of micro-subject is the emergence of "factory". For the formation of modern factories, both Karl Marx in *Capital*, and Paul Mantoux in *The Industrial Revolution in the Eighteenth Century*, had made much penetrating analysis and elaboration, which will not be repeated here.

From the handcraft workshops in ancient times to the factories in the modern and contemporary times, there have been many profound changes, especially on the following three changes: the first is that in the aspect of technology and production tools, machine production replaces hand-tool production; the second is that in the aspect of economic relations, new classes and social relations are formed because of the implementation of hired labor; the third is that in the aspect of management of organization, the functions of management and its ranks begin to stand out unprecedentedly with the increasing factory scales and more and more complex management, thus in the new situation and condition, the revolutionary change of management occurs inevitably. In general, it can be held that the emergence of factories means the co-appearance of the machine revolution, social relation revolution and management revolution.

For the evolution of the organizational form of the microscopic subject of engineering activities, the emergence of factories does not mean the end of evolution. On the contrary, under the conditions of modern economy, society and engineering, the concrete form of the "micro-subject" of engineering activities inevitably evolves in a faster speed, more complex and changeable form, and more profound "connotation".

---

[4]In order to simplify the problem, this article does not discuss engineering activities like the pyramids and the Great Wall that were organized by the "state" and the "ruler" in ancient times.

In the modern society, enterprises have become the main micro-subjects engaged in engineering activities, and enterprises have become an important research object in economics, management, and other disciplines, and even form a special research field—"Enterprise Theory". There have been various interpretations and analysis of the theoretical views of schools of the firm—the corporate theory of the complete contract framework, the transaction cost economics, the new property rights school, the theory of enterprise competence, etc. It is especially noteworthy that works specializing in studying enterprise evolution are still emerging frequently, such as *The Theory of the Growth of the Firm* (Peng, 2007), *The Evolutionary Theory of Enterprise Development* (Wu, 2004), *Research on the Evolutions of Economic Organizations* (Gao, 2009), *The Changes of the Division of Labor, Technology and Production Organization* (Xie, 2005), etc., which fully shows the importance of enterprise evolutions. However, on the other hand, the evolutionary theory of enterprise is now so mulriple expression that there is no unanimous theoretical paradigm, which in turn reflects the complexity of enterprise-evolution research and the "preliminary nature" of the current research level.

Since enterprise evolution is an important and complex project still in the "initial stage of research", this article only points out its importance, and no more introduction, analysis, and discussion.

(3) The Meso Level of Engineering and Its Evolution

The evolution of engineering not only manifests itself at the micro level but also at the meso level. As already mentioned above, the meso level refers to "trade union", "engineer societies", "industries", "regional economies" and "industrial cluster". If to say that different trade unions and industries are "meso types" formed mainly in terms of "classification standards" like product types, technological properties, etc., then "regions" (such as "Silicon Valley", "Third Italy" in northern Italy, etc.) are the "meso type" formed mainly according to the concept of space and geography. Recently, the concept of industrial clusters has come to the fore, and its meaning and implication obviously belong to the category or type of "meso" mentioned here.

Though the term "meso" is uncommon either in academic writings or daily language, it does not mean that economists completely ignore the analysis and study on the meso level, because the trade unions, industries, and regions incarnating the meso level have already been the study object of many scholars. It is particularly noteworthy that in recent years, not only the "general" industrial issues, but also the issues of "industry evolution" have been an object concerned by many scholars, which is really thought-provoking. In recent years, the works directly studying the "evolution" of industries can be said to be soaring surprisingly. Directly based on industrial evolution as the basic theme and even as the title of the book, it has probably reached dozens of types in a short period of three or five years, such as *Industry Evolution and Enterprise Strategy*, *The Evolution of Industry Organizations: Theory and Practice*, *The Evolution of Industry Clusters and Study on Regional Economic Development, Industry Evolution, Collaborative Innovation, and Sustainable Growth of Private Enterprises* and so on.

In the process of engineering development and evolution, more than the micro-level development and evolution, the development and changes of micro-enterprises inevitably affect the development and evolution of "meso" in an over-level way. In turn, the evolutionary process of meso will certainly have a profound effect on the micro evolution.

In the process of engineering evolution, the complex relations between the micro evolution and meso evolution must be noted and analyzed. For example, in the period of the first industrial revolution, the new "factories" applying waterpower and new textile machines for production emerged, to which we must admit that they had achieved "revolutionary changes" at the micro level of engineering activities (from "shop system" to "factory system"). However, their products still belong to textiles, so they cannot be considered as creating a new "industry"—they merely achieved a micro-level revolutionary development in an ancient industry. But for the factories, manufacturing steam engines, of Matthew Boulton and James Watt, they are exactly creating a "new industry".

In Marxist political economics, the division of labor is divided into two categories: intra-organizational division of labor and social division of labor. The division of labor within the organization has been discussed above, and Adam Smith's main concern is just the division of labor within thc organization. Howcvcr, the division of labor in human society not only occurs at the micro level, but also occurs at a higher level and wider scope. If there is a division of labor at the trade, industry, and region level, it is the so-called "social division of labor".

The evolution history of an engineering project not only shows the history of enterprise evolution at the micro level, but also shows the evolutionary history of trades, industries, regions, and industrial clusters at the meso level. At the meso-level of engineering evolution history, people have seen the process of "emerging industries" flourished, gradually developed, and even sharply developed. Meanwhile, people also see the process in which the "outdated industries" declined or even died out one by one, including, of course, the dramatic evolutionary process where some industries sprang up as "emerging industries" and later withdrew from the historical stage of as "outdated industries".

It should be emphasized that the drama on the "evolution and decline of trade, industry, and region is "performing" continuously not only in history, but also in the world today. For example, in the 20th century, mankind has witnessed the rise of emerging industries such as aerospace, aircraft manufacturing, nuclear power, computers, and networks; in China, we have witnessed numerous cases of regional rise like "Pearl River Delta," "Yangtze River Delta," "Household Appliance Industry in Cixi, Zhejiang", and "Haining Leather Industry". In the meso evolution of engineering (the evolution of industry and industrial cluster, etc.), a lot of experience, lessons, laws, and influence are contained, which is impressive, moving, regrettable, sigh-provoking, thought-provoking, and introspection-provoking.

At present, China's research on the evolution of trade and industry just begins with a roaring developmental trend and a booming research situation, but no significant breakthrough is achieved in theoretical research. It can be expected that in the future engineering and academic development, people will surely continue to make newer and deeper progress on knowledge of the meso evolution of engineering.

(4) The Macro Level of Engineering and Its Evolution

When analyzing and researching engineering, no one can deny the importance of the macro level—national, international, and global. Because the engineering activities of different industries and regions in a country must form an integral whole, there must be a macro level or "scale" of "nation" above the meso level when analyzing and researching engineering problems.

It is reasonable to regard Christopher Columbus's voyage of "the discovery of America" as the beginning of "globalization". Karl Marx and Friedrich Engels said in *Communist Manifesto*: "The discovery of America, the rounding of the Cape, opened up fresh ground for the rising bourgeoisie." "Modern Industry has established the world market, for which the discovery of America paved the way. This market has given an immense development to commerce, to navigation, to communication by land. This development has, in turn, reacted on the extension of industry; and in proportion as industry, commerce, navigation, railways extended, in the same proportion the bourgeoisie developed" (Marx & Engels, 1988).

Certain technical means and their "engineering realizations" are the basic premise and foundation for the reason why human beings can survive in the "globalization" environment. Take the Columbus's voyage mentioned above as an example, if there are no certain technologies on shipbuilding and leading navigation in the vast oceans, Columbus's sailing is impossible. If human society is still at the productivity level of "self-sufficient", and labor productivity is very low without "sufficient surplus products" for "others on other continents" and "demands for remote products", then globalization is impossible and not needed. In fact, just under the conditions of modern technology, engineering, and economic, such as the laying of transoceanic telecommunication cables, the running of jet airliners, the realization of the Internet, the emergence of transnational corporations, etc., the seeming "boundless earth" before becomes a "global village".

Anthony Giddens, a well-known sociologist, said: "In recent years, globalization has become a hotly debated topic. Most people accept that there are important transformations occurring around us, but the extent to which it is valid to explain these as 'globalization' is contested." Some people summarize the different views on globalization into three schools: skeptics, hyper-globalists, and transitionists (Giddens, 2006).

This article does not intend to make a concrete analysis and comment on these different views, and only needs to point out that these different schools have no divergence in affirming the "existence" of the globalized phenomenon. Actually, they just disagree and dispute on "the degree of globalization" and how does "globalization" interact with "low-level social phenomena".

On the issue of engineering evolution, there are scant research about how to understand the evolution of the macro level—national, international, and global level—and its interrelations with the evolution of the other two levels, however, it is a very important problem whose importance is still increasing and deepening.

What needs to be emphasized here is that the so-called three levels—micro, meso, and macro are indeed necessary to be distinguished, but divided in a rough and not-absolute way. This is because in the division and "definition" of the three levels, only the micro level of individual and the macro level of globe are not divergent, while the "micro" level of "enterprise" and the "macro" level of "international" both can have different interpretations (in modern society, the "boundary" and "meaning" of an enterprise are often not clear and can be interpreted differently, and international relations can have many "bilateral" or "multilateral" explanations), and the "definition" or "scope" of the so-called "meso" is even more varied. But in any case, the "three-level" framework and division are still important for their theoretical and practical significance, which can help us understand and analyze many issues of engineering activities and evolutions.

(5) The Interaction and Evolution of Micro, Meso, and Macro Level

Although the above has inevitably involved different levels of interrelationships, there is still a need for some analysis and elaboration on this issue. No matter seen from the perspective of reality and theory, the three levels of engineering—micro, meso, and macro are interpenetrated, interacted, and interplayed with each other. It is impossible for "micro" to leave "meso" and "macro", and "meso" and "macro" cannot be separated from "micro". When analyzing and studying the issue of micro-level enterprise evolution, if it cannot be placed in the general context of industry and national development, in other words, if the micro problem of enterprise cannot be analyzed, investigated and studied in the meso and macro "overall environment", then the "pure micro" study cannot avoid being "misguided". On the other hand, in the analysis and study of the "meso" and "macro" issues, if "micro data" and "micro basis" are left, the grandiose discussion of the so-called research on "meso" and "macro" will inevitably be reduced to the study on "castles in the air" or "mirages on the sea", or even cannot be studied at all. Therefore, to analyze and study in the mutual relations, interactions, and interplay among "three levels" becomes a necessary requirement.

The interaction between the micro, meso, and macro levels of engineering is extremely complicated. When analyzing these relationship problems, it may be considered that the most critical point is the need to study the relationship between the following aspects. The first is the "cross-level" relations between "embedment" and "beyond". The term "embedment" is borrowed or "transplanted" from economic sociology. In 1985, Mark Granovetter published *Economic Action and Social Structure: The Problem of Embedment*, developing and deepening the term "embedment" proposed by Polanyi and making it a basic category of economic sociology. "The most important of Granovetter's works …, which came to operate as a catalyst for the emergence of new economic sociology and which is probably the most cited article in contemporary economic sociology." "Very importantly as well, is the fact that for many readers Granovetter's article opened up a whole new world of research" (Swedberg, 2003). Obviously, "embedment" can not only be taken as the basic concept of economic sociology, but also become the basic concept for analyzing and studying engineering and its evolution.

The essence of the "embedment" concept is to oppose the isolated research and the "atomized" research method, which requires that the research object into a "broader environment". The issue discussed in this article aims to require analyzing and studying by "embedding" the "micro subject" or "meso object" into a higher level. On the other hand, it must be realized that while the "micro subject" or "meso object" interacts with other subjects in this level, it will also bring the influence and effect "beyond this level". Therefore, the specific analysis and research on the complex relationship how to not only "embed" but also "transcend" has become an important issue.

The next is the mechanism problem of "upward" effect or "downward" role among different levels. If the focus or the key point of analysis is the cross-level study based on a certain "specific micro-subject" or "special meso-subject" when analyzing and researching the interrelations between "embedment" and "beyond", the so-called mechanism problem of "upward" effect or "downward" effect is to require the "general" analysis and study on the "cross-level" interactions among the three different levels of micro, meso and macro. Many issues that have been concerned by entrepreneurs and scholars, such as industrial clusters, industrial concentration, regional industrial innovation, industrial synergy and division, industrial upstream and downstream, industrial localization and globalization, are actually the problems in close relation to the upside and downside effects of macro, meso, and macro.

The third is the problems of the "stereoscopic network structures" and complex relations of micro, meso, and macro. If the point mentioned above is just to require analyzing and studying the "cross-level" interactions among the three different levels, then the third point is to require more comprehensive and more "integrated" analysis and research.

Not only should scholars analyze the "numerous subject networks at the same level", but also analyze and study the relations of "multi-level stereoscopic networks". Among many complex relations involved, the relations between various positive and negative feedbacks of multiple "routes" is particularly noteworthy. When analyzing and studying the interactions among micro, meso, and macro levels in engineering activities, not only their various structural and functional problems, but also their dynamic and developmental problems are necessary to be analyzed and studied, in other words, to analyze and study their evolutionary issues. If engineering topics cannot be placed in the analytical framework and research perspective under "the interaction and evolution of micro, meso, and macro" when analyzing and recognizing engineering problems, the occurrence of some sort of defects or even errors are difficult to avoid.

6.5 Philosophy of Engineering as a Branch of Philosophy and Engineering Studies as an Interdisciplinary and Multidisciplinary Research Field

If it can be recognized that science, technology, and engineering are three different objects, then theoretically and logically speaking, the problems of the theoretical premises and foundation for the existence of three different branches of philosophy —philosophy of science, philosophy of technology, and philosophy of engineering —are ameliorated up to a half. Similarly, theoretical premises or foundation for that there may exist three different interdisciplinary or multidisciplinary areas of research—science studies, technology studies, and engineering studies—are ameliorated by a half.

From the history of disciplinary development, modern philosophy of science and modern philosophy of technology have a long history of development, and philosophy of engineering is only a baby.

In the 20th century, the philosophy of science has developed into a branch of philosophical research with fruitful achievements and tremendous influence. To some people, philosophy of science has become "postmature" so that Paul Feyerabend claims that the philosophy of science has become "a subject of great past" (Feyerabend, 1989). Feyerabend's remarks reflect his consistent language style. Although this remark is not groundless, many people still believe that the future of philosophy of science is bright, so do I.

The philosophy of technology has also came a long way and achieved fruitful results. Although the recognized origins of modern philosophy of technology can be traced back to the *Grundlinien einer Philosophie der Technik*, which was published by Ernst Kapp in 1877, nearly half a century earlier than the "Vienna Circle", it should be acknowledged that the modern philosophy of technology has seen a rough journey more than a century and has gained influence as well as academic achievements and that are not outshone by modern philosophy of technology. At present, some scholars in the United States are promoting the "empirical turn" of philosophy of technology. I believe the future of philosophy of technology is also bright.

Compared with philosophy of science and philosophy of technology, philosophy of engineering is still a baby in her fast-growing period. Obviously, the theoretical and practical potential for philosophy of engineering is extremely broad. I believe its future is limitless.

Both modern philosophy of science and philosophy of technology were created by foreign scholars. In establishing the philosophy of engineering, however, European and American scholars did not walk in front of Chinese scholars. Louis L. Bucciarelli from the United States published *Engineering Philosophy* in Europe in 2003 (Bucciarelli, 2003), while I published *Introduction to the Theory of the Artificial* in 1988 (Li, 1988) and published *An Introduction to Philosophy of Engineering* in 2002 (Li, 2002a), Changfu Xu (徐长福) published *Theoretical Thinking and Engineering Thinking* in 2002 (Xu, 2002). The publication of the above books may be regarded as the signal and symbol of the "launching" of philosophy of engineering in China as well as in Europe and the United States. Yongxiang Lu (路甬祥), president of the Chinese Academy of Sciences, and Chen Changshu, philosopher of technology in China, believe that the book *An Introduction to Philosophy of Engineering* is a ground breaking and original book (Chen & Bei, 2002; Lu, 2002). The preface, written by the famous philosopher Qinghai Gao (高清海), in the book *Theoretical Thinking and Engineering Thinking*, holds that the question raised in this book "is universal and even global and historic". At present, the development momentum of philosophy of engineering in China is much stronger than in the West to a certain extent. This paper does not intend to analyze why the construction of philosophy of engineering has failed to start earlier than in China and why the "momentum" of development is not sufficient in the western philosophical circle. This paper only wants to point out that western scholar "had a head start" in establishing philosophy of science and philosophy of technology while "blearily" failed to start earlier to establish philosophy of engineering.

Philosophy of engineering is not a game in the ivory tower. Its soul is the connection between theory and reality. The reason why China can walk in front of European and American scholars in the development of philosophy of engineering is that it has its profound social foundation and social reasons. At present, philosophy of engineering in China has been developing rapidly. And we hope that we can achieve more fruitful outcomes in this field in the future. For engineering practice and engineering activities, we not only need to consider it as the object of philosophical analysis and research, but also as the object of interdisciplinary and multidisciplinary research. That is to say, we not only need to establish a new discipline of philosophy of engineering, but also need to establish "engineering studies" as an interdisciplinary, multidisciplinary field of research.

Since the late 1960s, STS has emerged as an interdisciplinary and multidisciplinary field of research in the United States. STS can be an acronym for science, technology, and society, or the abbreviation of science and technology studies. Recently, the Chinese version of the famous *Handbook of Science and Technology Studies* has been published by Beijing Institute of Technology Press. In English, people can split "science and technology studies" into two parts, so there emerged two areas, science studies and technology studies.

Currently, in the West, science studies and technology studies are not only commonly used terms, but also have a "specialized" magazine—*Journal of Technology Studies* is a magazine published in the United States, while *Journal of Science Studies* is funded by Finnish Academy of Sciences. For more than 30 years, Western scholars have made great achievements in interdisciplinary and multidisciplinary research (STS or S & TS) on science and technology. The STS study has also had a great influence in China and has achieved reasonable success. However, as far as we know, there is no magazine in the West that uses "engineering studies" as the "theme." Although there are scholars in the West, such as Gary L. Downey and Juan C. Lucena, now consider it necessary to distinguish between engineering studies and technology studies, their claim seems to stand alone in that if we realize that the "Engineering Studies" is just one section of the book *Science and Technology Studies* and the chapter where the session is in is also simply named "Science and Technology Culture" (Downey & Lucena, 1995) which makes it "overwhelmed" like an "off-key" note in the powerful "chorus" of science and technology studies.

In conclusion, probably because many western scholars do not clearly differentiate between engineering and technology, it seems that we can still determine that Western scholars have not yet established "engineering studies" as an "independent" interdisciplinary and multidisciplinary field of research, according to the standards of sociology of science. Nevertheless, establishing "engineering studies" as an "independent" interdisciplinary and multidisciplinary field of study is a logical reasoning of "the triad of science, technology and engineering". Above all, the reality in China's engineering construction appeal to us strongly that not only must we conduct interdisciplinary and multidisciplinary research on scientific and technological activities, and be in need of these two interdisciplinary and multidisciplinary fields of research, we also need to conduct multidisciplinary and interdisciplinary research on engineering activities and establish "engineering studies" as an independent interdisciplinary and multidisciplinary field of research.

Recently, the Engineering and Social Research Center of the Graduate School of Chinese Academy of Sciences has started to publish *Annals of Engineering Studies*, and we hope this will help to establish "engineering studies" as an interdisciplinary and multidisciplinary research field.

Philosophy of engineering and "engineering studies" promote and permeate each other. Imre Lakatos, a famous philosopher of science, said, "Thus any historical study must be preceded by a heuristic study: history of science without philosophy of science is blind" (Lakatos, 1980). This Lakatos' Kantian-like viewpoint is not only suitable for us to understand the relationship between philosophy of science and the history of science, but also appropriate for us to understand the relationship between philosophy of engineering and "Engineering Studies" as an interdisciplinary and multidisciplinary field of research.

At present, people in China are making every effort to build a well-off society in an all-round way. We have carried out unprecedented engineering construction. In the future, we will carry out construction of a larger scale in new evaluation standards. In such social conditions and social environment, establishing the philosophy of engineering as a new branch of philosophy and establishing "engineering studies" as an interdisciplinary and multidisciplinary field of research has undoubtedly important theoretical as well as practical significance.

References

Agassi, J. (1988). The confusion between science and technology in the standard philosophies of science. In F. Rapp (Ed.), *Studies in the structure of thinking in the technological sciences* (pp. 51–52). Changchun: Jilin People's Publishing House.

Bucciarelli, L. L. (2003). *Engineering philosophy*. Delft: Delft University Press.

Bunge, M. (1988). Technology as applied science. Contributions to a philosophy of technology. Rapp, F. (Ed.). *Studies in the structure of thinking in the technological sciences* (pp. 28–50). Changchun: Jilin People's Publishing House.

Chen, C. (2002a). *Chen Changshu's collected works on philosophy of technology* (pp. 9–14). Shenyang: Northeastern University Press.

Chen, C. (2002b). Attaching importance to engineering, engineering & technology, and engineers. In Z. Liu & X. Wang (Eds.), *A Yearbook on the Study of Chinese Philosophy of Technology in 2001* (Vol. 30, pp. 28–29). Dalian: Dalian University of Technology Press.

Chen, C., & Bei, C. (2002). Creating a new frontier of philosophical research, Commentary on "*Introduction to Engineering Philosophy*". *Philosophical Researches, 10*, 73–74.

Downey, G. L., & Lucena, C. (1995). Engineering studies. In S. Sasanoff, et al. (Eds.), *Handbook of science and technology studies* (pp. 167–188). Sage Publications, Inc.

Enderle, G. (2002) Business ethics. In N. Bunnin & E. P. Tsui-James (Eds.), *The Blackwell companion to philosophy* (2nd ed., p. 536). Malden: Blackwell Publishing.

Feyerabend, P. (1989). Philosophy of science—There is a great past discipline. *Philosophical Translation*, (1).

Gao, Z. (2009). *Economic organization evolution research*. Shanghai: Shanghai University of Finance and Economics Press.

Greenwald, D. (1992). *Encyclopedia of economics* (T. Li, et al., Trans.). Beijing: China Social Science Press: 764, 287.

Giddens, A. (2006). *Sociology* (5th ed., p. 60). Cambridge: Polity Press.

In recent years, papers in this field, e.g. Bocong Li, (2000). The Distinction and Connection between Science and Technology, In Chuanjie Guo (ed.), *Scientific and Technological Innovation and National Rejuvenation,* Beijing: Xuexi Publishing House: 128–132; Wulun Jin (2000), The Boundary between Science and Technology Must Be Drawn. *Science and Technology Daily*. 15 December; Huaxia Zhang & Zhilin Zhang, (2001), The Research Program of Philosophy of Technology—From the Point of View of Demarcation between Science & Technology, *Studies in Dialectics of Nature*, 17(2):31–36; Xingmin Li, (2002), A Few Questions about Scientific Theory, *Social Sciences in China*, (1):21–23; Shixu Guan, et al. (2002). Further on the Relationship between Technology and Science in the Perspective of Innovation, In: Zeyuan Liu & Xukun Wang. (Eds.) (2002), *A Yearbook on the Study of Chinese Philosophy of Technology in 2001*. Dalian: Dalian University of Technology Press.

Lakatos, I. (1980). *The methodology of scientific research programme: Volume 1: Philosophical papers* (p. 53). Cambridge: Cambridge University Press.

Li, B. (1988). *Outline of artificial theory*. Xi'an: Shanxi Science and Technology Press.

Li, B. (2002). *Introduction to engineering philosophy*. Zhengzhou: Daxiang Publishing House.

Li, X. (2002). A few questions about scientific theory. *Social Sciences in China, 1*, 21.

Lu, Y. (2002). *Introduction to philosophy of engineering (preface)*. Zhengzhou: Daxiang Publishing House.

Li, Z., & Liu, Z. (2003). A review of Hans Poser's thoughts on philosophy of technology. In Z. Liu & X. Wang (Eds.), *A yearbook on the study of chinese philosophy of technology in 2002* (p. 247). Dalian: Dalian University of Technology Press.

Marx, K. (1992). *Capital volume I* (p. 461). London: Penguin Books.

Marx, K., & Engels, F. (1988). *The economic and philosophic manuscripts of 1844* and *the communist manifesto* (pp. 210, 211). New York: Prometheus Books.

Marx, K., & Engels, F. (1998). *The German ideology* (p. 38). New York: Prometheus Books.

Peng, L. (2007). *Business growth theory*. 2007. Shanghai: SDX Joint Publishing Company, Shanghai People's Publishing House.

Poser, H. (1998). On structural differences between science and engineering, *Techné* 4:3(winter), DLA Ejournal Home.

Poser, H., Liu, Z., & Li, W. (2003). Dialogue between Chinese and German scholars on technology and philosophy. In Z. Liu & X. Wang (Eds.), *A yearbook on the study of Chinese philosophy of technology in 2002* (p. 195). Dalian: Dalian University of Technology Press.

Wen, R. (1988). *Kaogongji guide* (p. 216). Chengdu: Bashu Publishing House.

Smith, A. (2003). *The wealth of nations* (p. 9). New York: Bantam Dell.

Swedberg, R. (2003). *Principles of economic sociology* (p. 35). Princeton: Princeton University Press.

Wu, D. (1996). *Wu Dayou's collected works on philosophy of science*. Beijing: Social Science Academic Press.

Wu, G. (2004). *Evolutionary theory of enterprise development*. Shanghai: Shanghai University of Finance and Economics Press.

Xie, F. (2005). *Division of labor, technology and production organization changes* (pp. 152–153). Beijing: Economic Science Publishing House.

Xu, C. (2002). *Theoretical thinking and engineering thinking*. Shanghai: Shanghai People's Publishing House.

Zhang, P. (1997). *The emergence and development of microeconomics* (p. 1). Changsha: Hunan People's Publishing House.

Part II
Contemporary Work

Chapter 7
Aesthetic Factors in Engineering Technical Activities

Baowei Zhu

Abstract It is the inherent requirement for acting and constructing based on the principle of aesthetics in working and other human activities. In modern times the engineering and artistic activities interact with each other in terms of ideas and methods, while they mutually establish the means for the other. Accordingly the two fields are promoted to be developed respectively, and in the meantime a distinctive aesthetics appears, which is distinguishable from those in the artwork and in the technical products per se. The aesthetics in engineering is in compatible with the practical function. It involves directly in the material, structure, craft and functions, and displays the harmony between human-machine and artificial products-nature. The aesthetic factors exist throughout every step of engineering activities, including design, construction, and use. Meanwhile, as internal criteria they generally affect the engineering processes and principles. As a trend, modern engineering tends to meet the abundant spiritual needs in additional to seek to achieve perfect functions. The Chinese ancient crafts have their own unique aesthetic system. They pursue a coordinated and reasonable arrangement of both nature and human life, and can be briefly characterized into two categories: "Harmony" and "Suitability". "Harmony" lies in the coordination, complementation and mutual excitation of heterogeneous elements, which is behind the principle of occurrence, the normal state of existence and the optimistic of functions. The Harmony is embodied between *Dao* and instrument, and *Dao* and crafts. "Suitability" is the degree of the Harmony. It involves in multiple relationships between such as "craft" and "material", people and objects, heart and objects. Suitable has intimated the ergonomics of present-day. This aesthetic system also comprises specific aesthetic concepts such as illusion and realism, rhythm, scene, elegance, subtleness and flatness. These not only contribute to the glory of Chinese ancient crafts, but also provide references to modern engineering technology aesthetics and industrial design. The problems brought about by the expansion of technological rationality can also be dispelled and counterpoised accordingly.

B. Zhu (✉)
Institute of Philosophy, Chinese Academy of Social Science,
No. 5, Jianguomen Inner Street, Beijing, China
e-mail: zhubw1949@163.com

Q. Wang (ed.), *Chinese Philosophy of Technology*, Philosophy of Engineering and Technology 34, https://doi.org/10.1007/978-981-15-1952-9_7

Keywords Aesthetic · Technology · Engineering · Harmony · Suitable

Today we live in a world built by engineering technology and its increasing products. In addition to bringing about benefits, convenience, comfort, increased capacity and powering growth, engineering technology has also participated in and undertaken the creation of a more beautiful life and a more beautiful environment. "It is constructed according to the law of beauty" has become an important principle in engineering activities.

7.1 Engineering, Art and Beauty

Things with the attributes of conciseness, harmony, completion, ingenious design and producing are regarded as beautiful, and they represent the embodiment of the high degree of unity of conformity of law and conformity of aim in human practice. Beauty brings pleasure to people, and its essence is freedom. Ancient Greek philosophers regarded beauty as the nature of the universe itself, while philosophers in modern times believe that aesthetic activities are at the "root" of human life, where human can feel the intimacy between themselves and the world. The essence of human activity is creative and free. Human's creation is fully and freely carried out in the labor i.e. practice of creation. On the one hand, objective factors of labor, such as object, environment and conditions, interact with people to form a harmonious perceptual unity, where the labor process itself forms a harmonious rhythm, and consequently people get the aesthetic experience; on the other hand, the externalization, materialization and objectification of human intelligence and physical strength become artifact-product. In turn, people can intuitively see their own essential strength from labor and their products, appreciating their own creation. In labor and other human activities, it is an inherent requirement to act and create based on the laws of beauty.

The reason is that if the conflicts between the law (truth) and the aim (good) cannot be coordinated aesthetically, in other words, if the harmony and unity between the law and the aim cannot be achieved, people cannot succeed in the production and life, thus failing to construct themselves. Moreover, the people's aesthetic concept, aesthetic consciousness and aesthetic ability are also formed and developed in the practice activities, in the productive labor, in the enlightenment and edification of the social and historical culture.

In modern times, art is regarded as the bearer of aesthetic function. In the early stage of human development, technique and art were originally integrated, and aesthetics was not completely separated from "skills". The ancient Greek "Τέχυη" is the general term for production activities, which includes both technique and art. The ancient techniques were based on producers' experience and intuition, and the mastery of scale relations, proportions and rhythms became the center of labor experience and technical improvement. The development of technology relied on the skills of using tools with both hands and also embodied, transformed and materialized human's feeling, spirit and personality, thus endowing the practical utensils with an aesthetic effect. The Neolithic pottery, the bronze ware of the Shang and Zhou period in China, the ancient Egyptian Pyramid, and in the architecture in the ancient Greece, the ancient Rome and the ancient China present the strongly perceptible beauty. As time progressed, these implements and buildings lost their initial functions, and were appreciated as works of art. A large number of ancient works of art seen in museums today were all practical implements at that time.

The industrial revolution replaced handicraft production with machine production. In the early stage of industrialization, industrial products were separated from aesthetics with a serious tendency of inhumanity in the relationship between people and machines. In 1851 at Britain's "World Industrial Expo", people were dissatisfied with the rough and ugly industrial products, as well as their tedious and vulgar decoration, and began a long process of seeking the unification between engineering technology and art, as well as the unification of using function and aesthetic in the large mechanized production. Initially, some artists were determined to get involved in industrial production, so "disinterested" art would return to social production and life. This resulted in the British "Arts and Crafts Movement". The advocators of the "Arts and Crafts Movement" proposed the use of artisans' handicraft production to overcome the malady of ugly industrial products, which had no obvious solution. But this movement impelled the combination of art and industrial production objectively, which integrated aesthetic factors into engineering technology design and influenced the evaluation standard of the product, and eventually led to the appearance of "engineering design" and "engineering aesthetics". By the twentieth century, the development of "industrial design" had become an important field in modern manufacture, and the artistic design work became an indispensable part of the factors in production and life.

In this respect, the Bauhaus movement played an important role. Bauhaus was originally the name of a school of Arts and Crafts established in Weimar Germany in 1919, and its ideal goal was to cultivate a group of builders of future society who would fully recognize the trend and needs of the industrial era in the twentieth century and have full power to use all scientific, intellectual and aesthetic resources to create a new environment that would meet both human spiritual and material needs. The Bauhaus movement demanded the abandonment of traditional forms and the extra decoration of the product in history. Their theoretical principle involved the form's following and embodying function, respecting the logic of the product structure itself, attaching importance to mechanical technology, promoting standardization and considering commercial factors. These principles are called Functionalism. Functionalists object to regarding beauty in products as additional "decorations" because they look at the decisive foundation of aesthetic form in products from the rationality of industrial production, and advocate that the aesthetic characteristics of products should be embodied in the purpose of engineering. Walter Gropius, the founder of Bauhaus, the first president and German modernist architecture master, also advocated the creation of the "visual language" of industrial design.[1] After years of effort, Bauhaus had formed a concise aesthetic style suitable for large-scale mechanized production, which raised the design of modern industrial products to a new level. Core elements of the Bauhaus movement worked hard to spread Bauhaus's thought and spirit, which had a great impact on the field of global architecture and industrial product design.

On the other hand, the development of engineering technology has greatly expanded the aesthetic field of people and changed people's aesthetic sense. Since the middle of the nineteenth century, projects and buildings seem to have become purely utilitarian or functional, and people have seen a different kind of beauty from traditional works of art. For example, the iron bridge of Thomas Telford, the Crystal Palace of the British Industrial Expo, and the Eiffel Tower in Paris represent the beauty of engineering and technology products, as opposed to that of traditional art. The speed, order, strength and even standardization in machine industry had become an important aesthetic standard in the new era. Correspondingly, there is the full penetration of industry and technology into the arts field. It has not only profoundly changed the expression of art, the aesthetic orientation and value orientation of art, but also changed the art itself, promoting the transformation of modern art in the twentieth century.

Thus, engineering, art and aesthetics are different types of human activities. All of them have different laws, logic, intrinsic values and social functions, and above all, methodological principles. Engineering is a "creation". The products it produces are material entities, satisfying the practical needs of people in the practical range; Art is spiritual creation. Its product is a "form" which satisfies human spiritual and emotional needs. Engineering technology pursues benefit, while art pursues beauty, so that since Immanuel Kant, people have generally used "practical value" to distinguish engineering technology products and works of art, "actual benefits" as the criteria for evaluating engineering technology, and "aesthetic pleasure" as an evaluation criterion of art. Engineering and technological activities must follow the laws of science, economy and society strictly, which is a rational process. Programming and standardization are also important features and methods of engineering technology, while art and aesthetics are emotional and personalized. The influences of engineering technology and art on the economy, society, natural environment and even human beings are totally different. Both engineering and art activities have their own boundaries and limitations. They are independent of another, yet complementary to each other, and constitute a full and colorful human activity, together with science and other forms. Engineering activities and art activities permeate and introduce each other in ideas and methods (also materials for art) , as a result, it impels the formation of "modern art" and shapes the appearance of today's engineering technology products and our world.

---

[1]"Visual language" is a special form of language that makes the thought have visual representation. Symmetry, visual illusion, color, light and shadow of objects, relation between space and volume, and relation between color and scale are the features of visual language.

It can also be seen that there exist two relationships between engineering activity and aesthetics: one is the relationship between engineering technology and art; another is the relationship between engineering technology and aesthetics. The relationship between engineering and aesthetics is often reflected in the relationship between engineering and art, as art is viewed as the concentrated (or even the only) embodiment of beauty. However, the relationship between engineering technology and aesthetics is not only coincident with the relationship between engineering technology and art, but also different from the relationship between engineering technology and art. Not all through the medium of art, engineering technology and beauty have their own internal relations which are different from art. Not only in artistic activities, the pursuit of beauty in every activity of human life and practice is determined by the essence of life and practice, and it is also the ontological basis of the beauty of engineering technology. The beauty of engineering technology has its own special nature, and its scope can be even wider than art. The beauty of engineering technology is directly based on function, structure, material and technology. It is a united beauty with practical functions. The harmony in engineering technology also includes the harmony between human and machine and the harmony between engineering technology products (artificial objects) and the natural environment, which is also different from the beauty of art.

7.2 Aesthetic Factors Throughout the Whole Process of Engineering Technological Activities

Engineering technology, art and beauty are all the products of human creativity, all of which serve human beings and are the highest goals of human activities—freedom and development. They also have similarities in ways of thinking and methods, such as using imagination and intuition, etc.

Generally speaking, the completion of a project or the advent of a technical product must follow these three stages: Designing, building and using. The aesthetic factor runs through the whole process.

The first stage is design. A good product is created according to a good design. The research (and practice) of the beauty of engineering technology is also mainly in the design of engineering technology (from different angles, people also call it industrial design, including industrial aesthetics, technical aesthetics, etc.). It involves aesthetic influence in three areas. The aesthetic elements should be introduced into the design. The purpose is to optimize and beautify the various material products and environment formed by the means of industrial production. In addition to its beauty and practicability, its basic standards include "man-machine harmony" and "harmony with the environment". Besides that, there are two guiding roles: one is related to cognition; the other is similar to satisfying emotional needs. The factors related to cognition are mainly simplicity, harmony and balance (symmetry), etc. When evaluating a design, people often mention three criteria:

efficiency, economy and simplicity (Billington, 1983). To be specific, efficiency involves the physical resources used by the system, the economy refers to all aspects of the system cost, and simplicity relates to aesthetics.

The role of aesthetic factors in the design of engineering technology is similar to the role of aesthetic factors in scientific activities. There are a large number of such cases in the history of science. Scientists, especially physicists, try to combine aesthetics with cognition and use simplicity and symmetry as a cognitive standard or auxiliary standard in the field of science. Albert Einstein, Werner K. Heisenberg, Paul. A. M. Dirac and others have all discussed this issue, for example, Dirac thinks that the beauty of mathematics lies in its symmetry in the theory of physics, and its importance is even higher than the experiment itself. One of the most famous examples is James Maxwell's equation of electromagnetic theory. Even in experimental science, simplicity is also used as a guiding principle for research or an explanation for the natural order of things. For example, there are the top ten "most beautiful experiments" in chemistry, such as the experiments by Henry Cavendish, Marie Curie, Ernest Rutherford and Louis Pasteur. The aesthetic features involved are direct reasoning, ingenious experimental design, and concise design. They are called "the most beautiful" because the experiments are given value from aesthetic considerations rather than their cognitive value (Ball, 2005). This is because since the ancient Greek philosophers, people have formed a basic metaphysical concept, i.e., the universe is harmonious, and nature has a simple mathematical structure. Simplicity is not only beautiful, but also makes it easy for people to understand and grasp intuitively, especially when we are faced with complex problems.[2] Furthermore, in most scientific research and engineering design, there are many options to choose from. Among many correct solutions, the simpler one is always better.

Harmony is the overall connection of the various heterogeneous elements, the suitability yardstick of their coordination and fit, and also the expression of the regularity and order of their existence and movement. An important feature of engineering and engineering methodology is integration with various heterogeneous elements achieving overall functions or objectives. Here, harmony becomes a suited and exact standard.

The above discussion mainly focuses on the cognitive aspect. On the other hand, the pursuit of simplicity and harmony is also a process where emotional needs are satisfied. When we finish a design in a simple, straightforward, ingenious and elegant way, or when we see it in a finished product, we will be joyful. In addition, designers will consciously or unconsciously be guided by aesthetic interest. In architectural design, some aesthetic interest will ultimately affect the goal of the design, for example, the style is rugged or ornate, simple or elaborate, bright or implicit, delicate or unsophisticated, classical or fashionable, and so on.

[2]For example, in software design, because simple software systems are easily understood and more reliable than complex systems, simplicity as a target for software engineering design becomes a practical and feasible need.

The second stage is the construction (or labor) process, that is, to enjoy pleasure in the process of construction (or labor). In engineering activities, workers, including engineers and administrators, have also obtained aesthetic enjoyment. As Karl Marx said, "My work would be a free manifestation of life, hence an enjoyment of life" (Marx, 1975a, 1975b). The expression of the essential strength of human beings is first embodied in their productive labor including engineering design, construction and use, and the form of beauty such as rhythm, harmony and freedom. "Labor creates beauty" not only means that labor is the "foundation and source" of beauty, but there are also other aspects that arouse special pleasure, such as the feeling of action and labor, including balance, harmony and so on. "Dismember oxen by Pao Ding" written by Zhuang Zi (庄子) described the free state of the full control of objects and the "plenty of room for the blade of a knife", on the one hand, and praised the beauty of his action which "kept time with the rhythm of the Dance of the Mulberry Grove", on the other hand (The second aspect of "design" mentioned above actually refers to this special aesthetic enjoyment in mental labor).

The third stage in which the aesthetic factor penetrates is the use process of a product. In addition to practicality, engineering technology products also carry (or symbolize) multiple meanings and values, including aesthetic values. When facing the Grand Three Gorges Dam or the Qinghai-Tibet Railway zigzagging across the snowbound plateau, people will praise their magnificence. In addition to their functions of "road" and "dam", people can also feel the wisdom and hard work of the engineers and builders, the strength of the state and the creative power of science and technology, and feel the beauty of engineering technology. About this point, Le Corbusier, a representative figure of Bauhaus, has a brilliant expression: "Engineers unknown to the world at large, mechanics in shop and forge have conceived and constructed these formidable affairs that steamships are…, If we forget for a moment that a steamship is a machine for transport and look at it with a fresh eye, we shall feel that we are facing an important manifestation of temerity, of discipline, of harmony, of a beauty that is calm, vital and strong" (Corbusier, 1985). The human world is almost all constructed and covered by engineering technology today. In addition to the growth of benefit, convenience, ability and strength, the aesthetic pleasure that we can enjoy in our life, work and environment is also being increasingly provided by engineering activities and their products. Engineering technology has been involved in the creation of a beautiful life and a beautiful world, and beauty has also become a choice and evaluating standard for engineering technology products.

7.3 Construction Based on Aesthetic Law

More than 170 years ago, Karl Marx raised a famous proposition in his book *Economic and Philosophic Manuscripts of 1844*: "Man therefore also form objects in accordance with the laws of beauty" (Marx, 1975b, 277). In his perspective, mass machine industrial production is the externalization of human beings' essential power. The relationship between men and their productions is (or should be) an overall possession in terms of material, spiritual, utility and aesthetic aspects. Human beings' possession of man-made materials "should not be conceived merely in the sense of immediate, one-sided enjoyment, merely in the sense of possessing, of having. Man appropriates his comprehensive essence in a comprehensive manner, that is to say, as a whole man. Each of his human relations to the world—seeing, hearing, smelling, tasting, feeling, thinking, observing, experiencing, wanting, acting, loving—in short, all the organs of his individual being, like those organs which are directly social in their form, are in their objective orientation, or in their orientation to the object, the appropriation of the object, the appropriation of human reality. Their orientation to the object is the manifestation of the human reality" (Marx, 1975b, 299–300).

Karl Marx compared constructional work between human beings and animals, and claimed that "An animal forms objects only in accordance with the standard and the need of the species to which it belongs, whilst man knows how to produce in accordance with the standard of every species, and knows how to apply everywhere the inherent standard to the object" (Marx, 1975b, 277). In other words, men are able to master and apply every object's regularity in compatibility with their own survival purposes, consolidating regularity and intentionality into a perceptual and harmonious existence. In this way, men's freedom and development would be realized, and beauty created. This is what Karl Marx called "internal dimension", and the criterion of (generalized) beauty investigated in this article. This criterion should be applied to every aspect of survival activities, including engineering technical practices.

The purpose of applying the principle and methods of beauty into engineering technical activities is to optimize and beautify various material productions and environment produced in the process of these activities. It is basic measurement of application is comfort, beauty and practicability, which is manifested on the following three levels.

The first level is formal beauty, that is, the appearance, style, structure, color and material quality of production. It is designed to inspire people's emotions and imaginations through the overall presentation of merchandise (figure, shape, color, scale and rhythm) and its integral stability, order, uniformity, balance and harmony, all of which reflects what the merchandise stands for underneath. After all, appearance is the first thing that draws people's attention in contact with any engineering technical production and aesthetic judgment is always involved when it comes to estimating and selecting goods. Consequently, people's choices and buying inclination are largely dependent on primary impression, whether the production is attractive or not. Besides, it is also important for a harmonious relationship between engineering technical production and the natural environment.

However, it is not merely about formal beauty or decoration, as the overall beauty of engineering technical production is closely connected with its function. The priority of a fancy car is automobile utility, and it is fancy as a car. Its streamlined metal case is designed in accordance with aerodynamics and material science. On the other hand, as listed above, inner beauty is one of the basic elements of engineering technical aesthetic, which is exhibited by structure, function, craftwork and material. Structural beauty represents high quality, efficiency, utility, reliability and so on. Functional beauty stands for applicability, security and ease of operation. Function reflects purpose and is based on structure and material. The integral stability, order, uniformity, balance and harmony mentioned above are of vital importance to the comprehensive relations between all the parts or elements, and which are much more important than the formal aspect. The representation of a project is aimed to reach harmony both in its inner parts and in its relationship to the surrounding area, as well as to maximize function. The engineering technical aesthetic not only requires uniformity between its structures and function achieved through harmony and order, but also demands uniformity between utility and semantics. In other words, it aims to express the guidelines and meanings regarding material, construction principle and style through its appearance with the uniformity between content and form. Engineering technical aesthetic is a harmonious relation among inner structural beauty, functional beauty and formal beauty.[3]

Given the fact that any engineering technical product is utilized by and meant to serve man, it is important that the products are not only operated easily and safely, but also help to make people feel comfortable and concentrated at work. In this way, the users are able to handle any obstacles or accidents during work in time and to feel satisfied emotionally. One of the important aspects in engineering technical aesthetic is to reach a harmonious relation between products and users (including man and machine), products and the environment. After World War II, ergonomics developed as a practical discipline which investigates relations between man-made products and its users. It analyzes and locates the optimal man-machine relationship, through which it can provide the basis for design. Its purpose is to improve human beings' work efficiency while improving people's values at the same time.

Taking designing and construction of cabins in manned spacecraft as an example. As is well known, astronauts live in very limited and almost isolated hermetic cabins during long spaceflight, and they are influenced by environmental factors which can be neglected or barely exist in normal conditions on earth. Since astronauts play the most active and crucial role in space control system, it is important to protect them as complicated biological beings from any injures in extreme situations. These are all closely related to the security and reliability of the overall spacecraft system. How should we handle astronauts' living and working problems in such a limited cabin environment to make sure they feel safe and comfortable? How do we make best use of equipment in terms of utility and maintenance, and to arrange all instrumentations in a proper and coordinated fashion? Chinese designers have put forward three principles: utility, harmony and figurativeness. All equipment in the cabin should conform to the flight mission, and the structure "is determined by practical process ability and reflects the fact that it is a 'home'". The principles and methods of engineering technical aesthetic are

[3]For example, the windows and mouse developed from DOS to Microsoft.

exploited to make the cabin structure acceptable to human beings, and to inspire one's affections and positive feelings. For example, it is possible to create comfortable and spacious environments by arranging light, color, sound, appliance, style and rhythm in a natural and harmonious way. As a result, astronauts in working cabins could be as inspired as in the labs on the ground, and they should also feel at home when living in the cabins (Liu & Wang, 1995).

The three levels listed above are logically progressive; also, they are realized and created with the development of engineering technical practice and social conditions. The advancement of science, technology and engineering requires progress in spiritual and emotional aspects. Meanwhile, negative effects in industrial and technological development have put forward the issue of humanization in manufacturing and products, as well as the relationship between man-made products and natural environment. On the other hand, abundant material products, more choices and fierce market competition all demand higher standards for products utility, personality and aesthetic. All these contribute to people's higher request for engineering technical aesthetic. The advancement of technology and industry has made it easier to handle technical or engineering problems of products. Considering their potential in the manufacturing value chain, the factors of beauty and comfort have been of growing significance in terms of improving added value, enhancing company' strength and core-competitiveness, and upgrading industrial structure (Li, 1994). The tendency of engineering technology nowadays is to meet more needs on the basis of fulfilling perfect functions.

The most important characteristic of engineering technical methods is "integration", in which aesthetic and artistic approaches have become indispensable and have been of growing significance. Evidently, this is a research field to be further investigated.

7.4 Aesthetic in Traditional Chinese Crafts

Here we will briefly illustrate aesthetics in traditional Chinese crafts. Being a unique system, the traditional Chinese crafts' abundant ideological resources have enlightening and referential meaning for engineering technology aesthetics and industrial designs today.

In fact, as early as the first half of the Twentieth century, some scholars noticed the value of these ideological resources. For example, the founder of the basic modern design course, one of the most important members of the Bauhaus School, Johannes Itten, introduced traditional Chinese philosophy and aesthetic thoughts, and combined them with modern science and technology, creating and developing an "elementary course". The primary purpose of this course was to liberate creative capacity inside each student by teaching basic knowledge and material traits of color. J. Itten taught students the philosophy of Lao Zi (老子) and Zhuang Zi, frequently quoting remarks like "*Dao* is grand; The heaven is grand; The earth is grand; and mankind is also grand" "Human beings emulate the earth; The earth emulates heaven; The heavens emulate *Dao*, and *Dao* emulates Nature", and students were asked to learn to draw Chinese landscape painting, to observe and interpret the world with the philosophy of Lao Zi and Zhuang Zi. In the human body painting class, students were not required to be accurate from an anatomical perspective. Instead, they were asked to interpret the models' gestures in an out of the ordinary way. Eton also mastered the primary regularity of formal constitution in formative art from "*Dao Te Ching*", and cited Lao Zi's "it might be something that provides the value, but it is nothing that provides the utility" to help his students learn construction theory.

The characteristics of aesthetics in traditional Chinese crafts will be discussed in the following three aspects.

The aesthetics of traditional Chinese crafts has its own features. It has a long history, and the manufacturing and invention of silk and porcelain have enjoyed great fame for a long time, even ranking the highest in the world. Moreover, it is the continuous traditional aesthetics rooted in unique philosophical and aesthetic thought and a deep cultural background, which is the first characteristics of aesthetics in traditional Chinese crafts.

As early as the Neolithic Age, Chinese ancestors emphasized shape, color and texture when making tools. Many tools are not limited to practical usage. Instead, they also displayed beauty in terms of texture, scale, balance, symmetry and rhythm. The magnificence and rareness in the Xia Dynasty, the Shang Dynasty, the Zhou Dynasty; vigorousness and profoundness in the Han Dynasty; delicacy and beauty in the Wei Dynasty and the Jin Dynasty; luxuriance and gorgeousness in the Tang Dynasty; elegance and light in the Song Dynasty; wisdom and practicability in the Ming Dynasty. Different periods have different features. What's common is that they combined together in a historical chain. This intrinsic line of development runs through China's aesthetic spirit and is deeply rooted in Chinese philosophy and culture.

It is known that a great part of Chinese philosophies and cultures are rooted in *The Book of Changes*. The basic spirit of *The Book of Changes* is "live", in other words, "the changes is living process of living beings", which is the philosophy of "live" or life. *The Book of Changes* determines the nature of universe as "live", and "living is the grand virtue in the universe". In a certain sense, "*Dao*" in Chinese philosophy is the ontology of life, the greatest reform in universe and the serenest embodiment of all. It is a dynamic entirety rather than an entity, unable to be objected or ready-made, only illustrated in the diagram of "harmony between the heaven and human" or "connection between subject and object" and directly perceived through experiences intuitively.

Accordingly, traditional Chinese aesthetic expresses the feelings and apprehension of "*Dao*", and embodies them in the settlement of life. "*Dao*" is represented in "art" in a way that "art" endows "*Dao*" with image and life and "*Dao*" gives "art" depth and soul. As a result, traditional Chinese aesthetics is different from the Western one in that it emphasizes emotional feelings rather than cognition and imitation, which experiences and understands the movement and meaning of life. This kind of feeling occurs in the dimensions of visual, auditory, gustatory, tactile and spiritual senses, in which gustatory sense holds an important part. Some of the important notions in traditional Chinese aesthetic developed through the gustatory sense, such as "lingering charm" "overtone" and "taste". Besides, tactile sense is important too. Jade is associated with virtue and the porcelain in the Song Dynasty was praised as jade. Obviously, both jade and porcelain are associated with the tactile sense.

However, the feeling is not merely dependent on sensations or a part of the senses; rather it is the devotion with body and heart, and it is the coordination between sensations and spiritual apprehension. It aimed to achieve poetic comfort in spirit rather than satisfaction of the senses, and to liberate self to reach mental freedom. To go beyond oneself is more than transcendence in a strict sense, which is not enslaved by materials while living in the world. It is different form Faustian surpasses from exterior world and in pursuit of the infinite world. Its purpose is to realize spiritual sublimation in reality, and to reach mental peace and living settlement through harmony with everything in the universe. In consequence, beauty is living and connected with ordinary life in traditional Chinese aesthetics, upgrading from reality to "life itself" in order to achieve mental liberation. The ancient Chinese people also paid attention to comprehension and mental conception while creating art. Consequently, aesthetics in craftworks made by hand were of high proficiency: its value lies more in heart than in hand, and the stress is on mental wisdom and its operation, in other words, to "show one's own ingenuity".

The second characteristics of aesthetics in traditional Chinese crafts reflected as the basic principle of "harmony". The notions, principles and regulations in traditional Chinese aesthetics pay more attention to the coordination, penetrability and promotion between heterogeneous elements, such as "Yin—Yang" (阴阳) "form—spirit" "false—true". "Neutralization" enables a harmonious and healthy life, which is the meaning of beauty in the world. The highest boundary of "neutralization" is "harmony between heaven and human". The scale of "moderation" and "neutralization" is "appropriation" and "fitness". All of these run through traditional Chinese craftwork aesthetics. As the starting points of traditional Chinese craftwork design, "harmony between heaven and humanity" is not only expressed in the relationship between humanity and nature, but also in relations between "*Dao*" and "technique", and "*Dao*" and "art", etc.

The relationship between *Dao* and implement as well as *Dao* and technique has always been important propositions in Chinese philosophy. *Dao* is secluded and of nothingness, and it can only be realized, enriched and developed through substantiation into concrete implements and production processes, which is known as "*Dao* is embodied in implement". Conversely, an implement is the carrier of *Dao*. The only way to realize the meaning, value and function for implements is to embody *Dao* through integration with thoughts and ideas, implying the pursuit of metaphysics in traditional Chinese culture. Traditional Chinese craftworks aim to actualize functions of aesthetics and delight through certain interest and states expressed by implements and their configuration. In detail, it is social harmony and order in the social relationship with each other; it is the unity of humanity and heaven between man and nature; it is the uniformity between heart and material, appearance and essence, form and spirit, material and art, utility and beauty etc. Both the ancient Chinese architectural style and traditional manner of dress display ethical meaning and aesthetic value of implements that are carriers of *Dao*.

As the implement making activities of craftsman, techniques (技, "*Ji*") is the hybrid for the carriage and the developer of "*Dao*". Craftsmen were called "Bai Gong" (百工, a variety of craftsmen) in ancient China, and they belong to one of the six major occupations in society. Craftsmen were the creators of good lives and some distinguished technical inventors were called the equivalent of saints. In ancient China, people participating in craftwork design include craftsmen, refined scholars and even educated emperors, which endowed lots of craftworks with not only civilian aesthetic taste, but also Chinese philosophies and aesthetic spirit. These craftworks with great cultural taste took on the educational function of ethics and religion. The relationship between "*Dao*" and "technique" is interpenetrative, complementary and mutually supportive.

On one hand, "technique" is the path to "*Dao*". When some craft has reached its peak, it is only one step away from "*Dao*", not only the laws in heaven and earth, but also the status of "metaphysics": beauty and freedom. On the other hand, "*Dao* is the perfect state of technique"…advanced technique is the reflection of "*Dao*", and "*Dao*" can be further understood and experienced through the method of "technique". This is not only an application but also an "upgrade" for *Dao*: a dynamic process from abstraction and shallowness to concretion and richness.

The third characteristics of aesthetics in traditional Chinese crafts reflected as the basic feathers which can be summarized by "harmonious" and "appropriate".

As with all techniques in the world, ancient Chinese techniques first of all stress "practicality", insisting the top priority is to be practical. An ancient story states that "it's intelligent to make a crossbar at the end of a carriage, but stupid to make a kite." The story relates how Mozi (墨子) spent three years making a wooden kite eagle that got broken in only one day. When his disciples complemented him on the delicacy of his technique, Mozi said, "My technique is inferior to that of the crossbar makers. It takes them less than one day and only a short piece of wood to make the crossbar, which can transport things heavier than three thousand *jin* to a far-away place and can't be worn out for many years. Mozi was indeed wise on this" (Li, 1994). Opposing wicked craft is a fine tradition of making in ancient. However, when the ancient Chinese stressed practicality, they didn't simply emphasize "usefulness". The core of "practicality" lies in the "usefulness" that can achieve "harmony" and "appropriateness", so as to coordinate and reasonably position both nature and man's life. As early as *Kao Gong Ji*, we can find the idea of taking harmony as beauty, saying "Good timing, proper place, beautiful materials, and subtle technique, will make a perfect piece of work altogether" (Wen Renjun, 1993). In other words, making a fine product needs all of the four factors and their cooperation.

"Harmony" here is the coordination of heterogeneous things in nature and society, which is the optimal state of artistic dialectics, through which they mutually complement and enhance each other. "Harmony means conforming to others of different sounds." Generally speaking, harmony represents the supreme pursuit of the Chinese nation for an aesthetic state of the universe and life. It doesn't simply emphasize duality or conformity, but denotes the principle of things' occurrence, their normal state of existence and the excellent condition of their function. First of all, it means the best state of existence and the internal basis that gives birth to everything. This best state does not only belong to the whole, but also each factor in its best condition seen from its condition and function. The mutual dependence and development of all factors will yield something new and even better. That is "harmony will make all things grow and develop" (Wu et al., 1994) means. Such as "the coordination between the sour, sweet, bitter, spicy and salty makes delicious food," "the coordination of various sounds makes beautiful music," and "beauty is produced by arranging different things alternatively". Secondly, it means the elegance and peace of some artistic work, with profound artistic quality. Moreover, it means even-tempered and moral unity, or rather, it has such political and dogmatic functions as to coordinate social conflicts, improve human relationships, as well as adjust and balance spiritual and physical health.

While "appropriateness" is the extent to which "harmony" has been achieved (including the constraint of extreme desire). Most ancient Chinese works concerning technique emphasize "appropriateness", such as "each does as it suits" in farming, "delicacy lies in appropriate size" in garden layout, and "do as it is appropriate" in embroidery.

"Appropriate" first refers to the choice and use of materials. The thesis about timeliness, terrain, materials and technique in *Kao Gong Ji* first talks about the best time, condition and technique in choosing and using materials and their cooperation.

Excellent products should have beautiful materials. As far as the product's function is concerned, including aesthetic function, the material choice is significant. However, what matters more is "use". "Examining the curve and straightness of materials and making them into tableware according to their respective features", means to make full use of the nature, appropriate function and most beautiful aspect of materials. Take furniture making in the Ming Dynasty for instance. The choice of materials and workmanship were so excellent that such hard wood as rosewood and red sandalwood were chosen. Hard wood has a strong, fine quality, and beautiful texture. To make the beauty of materials stand out, the coating usually doesn't need to cover its texture and instead puts the side with the most beautiful texture in the most prominent place. The furniture made in this way is elegant in color, such is the effect that red sandalwood furniture has, with a similar luster to silk, whilst rosewood furniture looks like an amber.

The "appropriateness" in choosing materials also includes two important principles: materials should be made best use of and over development of resources should be prevented so as to keep the balance and reproductive ability of nature. As stressed in *Exploitation of the Works of Nature*, while exploring nature, we should develop and use it within certain limits so as to give it an opportunity to rest and recover and keep away from going beyond the limits of nature's endurance. Take the law of "pears' leaving and return" for an instance. It says that the number of pearls produced is limited. If they are exploited too often, the number of pearls will decrease. Only by stopping exploitation for decades and leaving the clams to reproduce themselves, can more pearls been produced. It actually depends on the increase and decrease principle of pearls. Also, this book stresses saving resources and waste recycling. Such as in paper production and silk reeling, "picking up the fragments of paper and making them into paper again, is called reviving paper" (Pan, 2008). This is the conduct of making full use of waste and indicates that using leftover materials saves resources and produces wealth as well.

Here is involved the relationship between "work" and "materials", or even between man and things, mind and objects. That is what is meant by "good timing, proper place, beautiful materials can be combined only by skillful craftsmen". The supreme state of the craftsmen's use of their wisdom and technique is "natural", or rather imitating nature in a natural way. "Although man makes it, it seems to be made by nature." Take furniture of the Ming Dynasty for instance. Its different parts are connected by various mortises and tenons without decoration. It has a brief and elegant shape without redundant varieties and mainly depends on straight lines supplemented by curves so as to achieve the variety of straight and curved lines and the combination of hardness and softness and impress people as intimate and mild. What is emphasized here is not "enslaving things" but not being enslaved by things, and in a way that desire is not set free either. That is what "appropriate" means.

The second principle of "appropriateness" is suitable to people or appropriate to people, or rather useful and beneficial to people.

First of all, people are attached great importance to. The value of utensils lies in their usefulness and they are aimed at facilitating people's life and satisfying their daily necessity. Mozi says the purpose of the magnanimous is to be found in procuring benefits for the world and eliminating its calamities (Shi, 1988). The idea shows that the aesthetic technique has been turned from being overelaborate and strange to everyday life and use.

Secondly, being beneficial to man means something that has everyday use for common folk to many people. Especially in the Ming Dynasty, Wang Gen, the initiator of the School of Taizhou (泰州学派), came up with the idea that the "everyday use of common folks is *Dao*." He considered everyday use of common folks as "the inherent reason", believing whatever has nothing to do with everyday use by common folks belongs to heresy (Gen Wang, 王艮). *Exploitation of Works of Nature* shows prominently the designing and creating thought that was given to people's livelihood and in the service of common people. Yingxing Song (宋应星) proposed at the very beginning of the book that the subject of the five cereals were arranged before the content on gold and jade because they more superior. In this way, he gives priority to serving the common people. In *Exploitation of Works of Nature* we can find that its content, arrangement and function etc. focus on the idea of designing and creating to serve lower class people so as to satisfy their daily needs.

Thirdly, "satisfying man's needs" does not mean "satisfy man's needs in a common way", but emphasizes suiting and pleasing man, to some extent similar to modern human engineering. Even in the Neolithic Period, the primitive people had been very selective about the scale and proportion of pottery. For example, the diameter of a basin's opening is always under 30 centimeters, less than the width of a man's shoulders to enable people to save efforts in holding it, and the diameter of a bowls' opening is often around 15 centimeters and the proportion of its diameter to height is generally 2 to 1, which makes them solid and stable and rather similar to those of today. In the Spring and Autumn Period, it became mainstream to stress suitableness. *Kao Gong Ji* requires the height of a wheel should be convenient for people to get on the carriage; the roof should not block the passenger's view; weapons should not be more than three times longer of a human body (Wen, 1993), and so on. These take "being suitable for people to use" as their starting point. To the West and East Han dynasties, there appeared many design and production methods giving priority to usefulness in decorating. The prominent example is the wild goose lamp. The lamp consists of the head and body of a wild goose with a fish in its mouth, two shades, and a fire pan with a crooked handle. All of these parts being able to be disassembled. The two shades can move freely, adjusting the direction of light and protecting the fire from being put out by wind. The belly of the goose can contain water and the smoke goes through the neck and into the water so as to reduce smoke pollution. Its subtle and delicate design is a masterpiece of ancient Chinese lamps. Besides, the heating chair, lamp candle and letters designed by Yu Li (李渔) in the Qing Dynasty in his *The Idle Feeling Is Sent Occasionally* can be taken as a model of something beneficial to the human body and that presents a humanistic idea of design.

"Appropriate to people" should consider different situations of different people specifically. In "Arch Men" of *Kao Gong Ji*, arches and arrows are not made according to the same specification but to different physique, appearance and characteristics of users. Different arches and arrows should be made for the stout and short with tolerant characteristics and slow actions and the strong-willed, short-tempered with swift actions (Wen, 1993). The fullest play of the function of arches and arrows depends not only on arches and arrows themselves but also on men, or rather the reasonable balance between men and things.

"Appropriate" or "suitable" also includes man's spiritual sustenance. The sustenance of life, to a great extent, refers to that of spirit. Zhaowen Wang (王朝闻), an aesthetician, says, "When it comes to the practical side of fine arts technique, we always substitute applicable for practical". For example, the shape of primitive pottery shows the spiritual need to adjust to aestheticism. Therefore, the spiritual applicable value of "fine arts technique" (in the words of Zhaowen Wang) is certain to surpass its useful value (Wang, 2007). All in all, being appropriate and suitable to people calls attention to caring about people's life, which should not be simply reduced to humanism but means the harmony between mind and body, heart and things, one thing and another so as to promote the healthy development of man and nature.

There are many special aesthetic categories in ancient Chinese techniques, such as asthenia and sthenia, charm, artistic conception, elegant and implicit, returning to plainness and so on. Since it's impossible to elaborate them one by one in this article, two of them will be dealt with here.

The first one is the mutual promotion of asthenia and sthenia. Asthenia refers to actual things, scenes and something described delicately while sthenia gives space for people to associate and imagine, where endless charm and artistic conception are included. In *Kao Gong Ji*, the idea of mutual promotion of asthenia and sthenia has been adopted in "Carpenter Making Frame of Musical Instruments". It says specifically that in sacrificial offerings, animals with fat will be used as the sacrificial animals; images of animals without fur, with feathers and scales are used on stand, carved with animals uttering sounds with different parts (Wen, 1993). This is about the production of musical instruments, producing the different parts by associating the images of various animals with the playing of music so as to achieve synesthesia. Baihua Zong (宗白华) points out its enlightenment upon the issue of asthenia and sthenia that the sound of a bell and chiming stone is already arousing beauty, but when making the stand for those instruments, the craftsmen are more than making a stand, since they need to design it as a complete and unified whole. Putting tigers and leopards under a drum enables people to visualize those animals while hearing the drum. The visualization and hearing are combined virtually in one's mind so that it's like the tigers and leopards are roaring. In this way, the carved animals become livelier and the sound of the drum becomes more expressive and charming and the whole artistic work is more touching and influential (Zong, 1981).

The mutual promotion of asthenia and sthenia triggers people's association and imagination which leads them to reflection and appreciation. Traditional Chinese aestheticism emphasizes artistic conception and charm, pursuing the connotation beyond images. Artistic conception is more often found in poetry and painting, and sometimes in technique too, where it makes its direct way into the design of gardens and even rooms. It is also shown in other techniques. For example, Guimeng Lu (陆龟蒙) in the Tang Dynasty writes in his poem that "in late autumn, the rough wind blows bleak, but when the Yue kiln opens, green mountains are seen" to praise the artistic conception of the blue china of the Yue kiln.

The second one is implicit and elegant. In traditional Chinese techniques, there are two pursuits developing side by side, one is the embellished beauty and other is light and elegant beauty (also described as the beauty of a lotus flower rising out of water) colorfully and dazzlingly. Being contradictory, the latter is superior to the former and can be considered as the elevation of the former or a state achieved by negation of negation, the ultimate goal of people's pursuit. This elegance and implicitness are most typically presented in the pottery of the Song Dynasty. The two Song dynasties are the golden age of Chinese pottery art, with a large number of famous kilns whose celadon wares have a bright and smooth texture and take quietness and implicitness to the extreme. The most prominent feature of Song pottery is that there is no decoration except some opening glaze of different sizes, integrating beauty into the enamel as soft and smooth as jade and brief and elegant in shape. The Ru kiln and official kiln both beat others with their smooth and delicate enamel and the shape lines are mostly curves with small radians, without sharp twists and turns, and decorative accessories, which make the shape of pottery powerful and straight. Most of the pottery has a plain surface and even if it is decorated, the decoration is just some tangling branches, plucked flowers interweaving, which shows an elegant and quiet style. Actually, it's not only the case with pottery, but also with textiles and gold and silver utensils. Although these classic works are exquisite, they have nothing to do with the showing off of techniques but instead pursue implicitness, naturalness, thickness within thinness, and strangeness within ordinariness which makes them approachable and triggers rich imaginings. This is the core of Chinese national culture.

It requires an elevation and detachment to shift from colorfully and dazzling embellishment to elegance and lightness. The exquisite technique and nature of things (or even mind) are connected perfectly. This involves the subtle use of wisdom and technique but this use also appears very natural and shows the perfect oneness of man and nature. Man's spirit has its quietness and purification in the aesthetic process.

From the above analysis, it can be concluded that the aesthetics of traditional Chinese craftsmanship, especially the profound philosophical and aesthetic ideas, are valuable cultural heritage. To inherit this heritage critically, explore it (including reflecting upon the internal reasons for its success and declination) and pursue a modern transformation will not only benefit today's aestheticism of engineering and technology and industrial design, but also will solve the problems and challenges brought about by the development of technology and expansion of technological rationality.

References

Ball, P. (2005). *Elegant solutions: Ten beautiful experiments in chemistry*. Royal Society of Chemistry.

Billington, D. P. (1983). *The tower and the bridge*. Princeton University Press.

Corbusier, L. (1985). *Towards a new architecture* (p. 91, 92, 102, 103). New York: Dover Publications.

Li, Y. (1994). *Translation and annotation of Han Fei Zi* (p. 414). China Bookstore Press.

Liu, S., & Wang, M. (1995). Engineering aesthetic designing of cabins in manned spacecraft. *China Aerospace*, 10.

Marx, K. (1975a). Comments on James Mill, Elémens d'économie politique. In K. Marx & F. Engels (Eds.) *Marx & Engels collected works, volume 3: Marx and Engels: 1843–1844* (p. 228). London: Lawrence & Wishart.

Marx, K. (1975b). Economic & philosophical manuscripts of 1844. In K. Marx & F. Engels (Eds.) *Marx & Engels collected works, volume 3: Marx and Engels: 1843–1844* (p. 277). London: Lawrence & Wishart.

Pan, J. (2008). *Translation and annotation of Tian Gong Kai Wu,* volume 2. Shanghai Ancient Books Press.

Shi, J. (Eds.) (1988). *Selected readings from famous Chinese philosophy with annotations and english translation, volume 1* (p. 74). China Renmin University Press.

Wang, G. *Complete works of Wang Xinzhai, volume 2*.

Wang, Z. (2007). Applicable and super practical. In *History of Chinese aesthetic technique* by Jian Hang (p. 3). People Fine Arts Publishing House.

Wen, R. (1993). *Translation and annotation of Kao Gong Ji* (p. 117). Shanghai: Shanghai Ancient Books Press.

Wu, G., et al. (1994). *Translation and annotation of Guoyu* (p. 488). Shanghai Ancient Books Press.

Zong, B. (1981). *Aesthetic walk* (pp. 32–33). Shanghai People's Publishing House.

Chapter 8
The Sinicization of the Philosophy of Technology and the Research on Socialization of Technology

Fan Chen and Jia Chen

Abstract The historical orientation and the process of Sinicization of philosophy of technology are important issues in the study of philosophy of technology in China. Firstly, this article reviews historical process of origin and development of philosophy of technology in China, and analyzes the theoretical origin, scientific and technological basis and organizational characteristics of Chinese philosophy of technology in its development path. Secondly, it explores the philosophical issues of technology concerned by the academic field of contemporary Chinese philosophy of technology, as well as the philosophical issues in the development of contemporary Chinese technology. Thirdly, this article introduces the basic arguments, standpoints and principles of the school of philosophy of technology with Chinese characteristics, as well as the approaches and paths of Chinese philosophy of technology in the dialogue between China and the West. Finally, it puts forward a theoretical exploration of the sinicization of philosophy of technology which is to establish a socialization theory of technology for integrating technology and society.

Keywords Philosophy of technology · Technology and society · Northeastern school · Introduction

Contemporary Western philosophy of technology began its process of institutionalization in 1975. After more than 40 years of development, contemporary philosophy of technology is in a period of multidimensional development. In the study of contemporary philosophy of technology, philosophers are faced with such practical issues as follows: How to view the relationship between philosophy of technology, philosophy of science and philosophy of engineering; how to deal with

F. Chen · J. Chen (✉)
Department of Philosophy, Northeastern University, No. 195, Innovation Road, Hunnan District, Shenyang, China
e-mail: neuchenjia@163.com

F. Chen
e-mail: chenfan336@sina.com

Q. Wang (ed.), *Chinese Philosophy of Technology*, Philosophy of Engineering and Technology 34, https://doi.org/10.1007/978-981-15-1952-9_8

the binary conflict between philosophy of technology in the engineering tradition and philosophy of technology in the humanities tradition, going beyond the contradiction between technological optimism and technological pessimism, opening the "black box" of technology in order to accurately deal with the composition and development stage of technology, the relationship between technology and society, technology and science, proposing a research program for philosophy of technology, and fully exploring new forms of philosophy of technology in contemporary development. To settle these issues, the Western philosophy of technology, on the basis of combining the two major research traditions, introduces the research method of constructivism in the sociology of scientific knowledge and the research method of context theory in philosophy of science, and gradually transfers from a traditional philosophy of technology to the contemporary philosophy of technology.

The turning of contemporary philosophy of technology also puts forward a new topic for the development of Chinese philosophy of technology. First of all, the development of contemporary philosophy of technology is not as stable and concentrated as the classical philosophy of technology in the mainstream approach, development context, and research interest, but it is characterized by various schools of thoughts and methods. Secondly, the development of contemporary philosophy of technology is changing rapidly, and shows highly integrated features on the basis of a high degree of differentiation. The distinctive practice orientation of philosophy of technology promotes its continuation of the traditional ethical concerns, and at the same time, the philosophical features of reflection have put forward modern philosophical inquiries in its foundation. This makes the contemporary philosophy of technology face severe challenges, and also provides broad prospects for the development of philosophy of technology. Thirdly, regarding the development of the Chinese philosophy of technology, since China's reforms and opening up, we have made better progress in tracing the trend of Western philosophy of technology. However, at the same time, we also face the difficult choice of "doing continuously" or "doing subsequently", especially in transcending the discourse system of western philosophy of technology to form a research program of philosophy of technology with Chinese characteristics, and a good dialogue mechanism between Chinese philosophy of technology and Western philosophy of technology.

8.1 The Sinicization Process of Philosophy of Technology

The study of modern Chinese philosophy of technology can be traced back to the forum on Dialectics of Nature held in Yan An in the 1938. The studies of Dialectics of Nature in this period guided the practical activities of science and technology by applying dialectics of materialism, contributing to the social progress of China. After the 1950s, according to the principle of serving the nation, Chinese studies of the Dialectics of Nature explored the laws of the development of science and technology, forming the development of institutionalization. Various universities have established research institutions for Dialectics of Nature, as well as running courses on the subject. In the Institute of Philosophy of the Chinese Academy of Sciences, a research group on Dialectics of Nature (1956), headed by Guangyuan Yu (于光远), was established. The *Journal of Research and Communication of Dialectics of Nature* was published by this group. It was the first journal on Dialectics of Nature in China.

As an important part of the Dialectics of Nature, the philosophical study of technology emerged at this time from the discussion on the methodology of engineering technology. After 60 years of continuous development and the unremitting efforts of several generations, the study of Chinese philosophy of technology is gradually becoming more mature. On the one hand, it was in the formation and perfection of its institutionalization. The philosophy of technology has gradually gained its legitimacy in the field of philosophy and social science and has become increasingly recognized. The research community of philosophy of technology is growing stronger. More and more scholars have started to show a strong interest in this field, thus increasing academic exchanges and communication. On the other hand, the topic area for its research is gradually expanding with increasing depth. The research methods are becoming more diversified and the research results more abundant. As early as 1947, the well-known Chinese scientist Xuesen Qian proposed the concept of "technological science". From 1955 to 1957, Xuesen Qian made a series of reports "On Technological Science" and published "Methodological Issues in the Study of Technological Science" (Qian, 1957), laying the theoretical foundation for the establishment of Chinese philosophy of technology. At the same time, Changshu Chen also published an article in the *Journal of Research and Communication of Dialectics of Nature*, calling attention to the methodological issues in technology (Chen, 1957), creating a theoretical precedent for the study of contemporary Chinese philosophy of technology. In 1960, Harbin Institute of Technology organized a discussion on the laws governing the internal contradictions of machine tools and organized a national symposium on Dialectics of Nature. After the symposium, they published an article entitled "On the Law of Contradictory Movements in Machine Tools from Block-type Machine Tools" in *Guangming Daily* (the text was later read by Chairman Zedong Mao (毛泽东) and reprinted in *Red Flag* magazine), which greatly triggered the interest of engineers and technicians in researching the issue of technical regularity, and this is also an important beginning for the study of the dialectic method of engineering technology in China (CPC General Branch Committee of Machine Tool and Automation, Department of Mechanical Engineering, Harbin Institute of Technology, 1960). During this period, a group of engineering and technical workers discussed and studied issues such as technological development, machine new materials and technological revolution.

From the early 1960s, the study of Dialectics of Nature initiated a discussion on the dialectics of production practice and technological development, which was an important supplement to the research planning of Dialectics of Nature at that time (Gong, 1991). The philosophical study on technology in this period had not yet formed its own unique disciplinary approach and research paradigm, and no institutional societies or journals had emerged. It can be said that during this period, "Chinese philosophy of technology research activities were basically in the spontaneous or semi-organized state" (Chen, 2005, 235), belonging to the embryonic stage of Chinese philosophy of technology.

After the reform and opening up, the studies of Dialectics of Nature obtained a good opportunity because its unique character of highlighting the combination of theory and practice enabled it to gain greater recognition. The study of Chinese philosophy of technology continued to develop based on Dialectics of Nature.

In 1980, Professor Changshu Chen proposed that "it is necessary to strengthen the study of the laws of technology and technology development from a holistic perspective". This is the first time that Chinese philosophy of technology as a specific research area was proposed. In November 1985, "The First National Academic Symposium on Technology Studies" was held at Chengdu University of Science and Technology. At that meeting, the Technological Symposium of Chinese Society for Dialectics of Nature was established (later changing to the "Professional Committee for Philosophy of Technology" at the second annual conference in Changsha in 1988), marking the formation and gradual maturation of the establishment of a philosophy of technology research with technology as the research subject. The Professional Committee for Philosophy of Technology also holds meetings every two years and has held 17 meetings so far. The biennial meeting also published a number of conference proceedings and established a website.

The 1980s was a period of independent and specialized research on philosophy of technology in China, and there was specific focus on technology as the research object. From the perspective of the theoretical study of philosophy of technology, a batch of theoretical monographs was published and a series of special studies on philosophy of technology were initiated. Most of the scholars of philosophy of technology have a strong engineering background. In terms of content and method, early Chinese philosophy of technology focused on the methodological study of engineering technology philosophy. Such as, Changshu Chen from the Northeastern University who launched the "Engineering Methodology Research"; Zeyuan Liu at Dalian University of Technology who conducted a "Study of Technology Development Methodology"; Shixu Guan at Harbin Institute of Technology who conducted "Study on the Structure and Development of Engineering Technology"; Derong Yang (杨德荣) from Chengdu University of Science and Technology who conducted the "Research of Japanese Technology Studies". The academies in universities of science, engineering, agriculture, medicine, and the military also conducted research on professional technology studies such as chemical technology studies, petroleum technology studies, agricultural technology studies, medical technology studies, and military technology studies. The philosophy of technology in this period was mainly combined with the analysis of history, economics and methodology of specific technology, coinciding with the historical origin of the philosophy of technology in Western countries. *Philosophy of Factory* and *Dictionary of Technology* published by the British economist Andrew Ure in the mid-19th century, and *History of Technology* published by the

German historian J. Popa can be regarded as the predecessors of Western philosophy of technology. The early philosophy of technology separated from the study of economics and economic history, turning the economic and historical direction of technological research to philosophical technological research (Chen & Yuan, 2001). The research orientation of the engineering tradition and engineering methodology in the early stage of Chinese contemporary philosophy of technology was initiated and promoted by Chinese philosophers of technology in a relatively independent situation, and a similar research path of philosophy of technology was formed.

Since the 1990s, the philosophy of technology in China has been introducing and commenting on the philosophy of technology from Western countries. The epistemology of technology, the axiology of technology and the applied research of philosophy of technology have all entered the field of philosophy of technology. Since the 21st century, Chinese philosophy of technology research has witnessed the largest number of articles published, the widest range of research fields and the stabilization of the academic team. Chinese philosophy of technology witnessed vigorous development in the new century.

8.2 Main Topics of Contemporary Chinese Philosophy of Technology

Due to the rapid development of Chinese philosophy of technology since the reform and opening up and the continuous influx of international philosophy of technology research themes, philosophy of technology research in China presents a gratifying scene, and at the same time demonstrates the controversies and different views on the focus of research topics among Chinese scholars, which is directly related to the study of Chinese philosophy of technology. Normalization and future direction of development are serious considerations for the future development of Chinese philosophy of technology (Chen & Chen, 2009).

1. **Coverage of topic fields in the study of Chinese philosophy of technology**

As a latecomer in the field of philosophy, since the birth of Chinese philosophy of technology in China, it has been characterized by a diversity of discipline paradigms and the complexity of its research methods. Many scholars have also pointed out the classification of the research field of the philosophy of technology. For example, Professor Wenhua Chen (陈文化) pointed out in the article "Review and Prospects of the Research on the New Philosophy of Technology in China", that philosophy of technology includes "ontology, axiology, methodology, sustainable development of natural technology, Marxist philosophy of technology, European and American philosophy of technology, and technological innovation and the transformation of natural scientific and technological achievements" (Chen, 2005, 236–237). Jianjun Zhao believed that the research fields of the philosophy of technology mainly include three basic dimensions, that is, basic theoretical research of technology, the application of technology, and the empirical research of technology. Lianghua Gao (高亮华) held that the study of philosophy of technology should include the nature of technology, the mechanisms and laws of technological development, technology and nature, technology and society, the epistemological and methodological issues of technology, and the application related to the philosophy of technology. It should be pointed out that the above thesis is that Chinese philosophy of technology has been intensively discussed for the past four decades, but does not mean that the scope of the topic area of philosophy of technology is limited to this.

The ontological research for Chinese philosophy of technology focuses on a series of issues such as artificial nature, the relationship between technology and science, technology and engineering, as well as the concept, essence, structure, elements, classification, skills and technological artifacts of technology. Regarding the issue of artificial nature, Changshu Chen believes that "artificial nature" is a basic concept of "natural remaking theory" and is a logical starting point of philosophy of technology (Chen, 1999). Based on the different understandings of the relationship between technology and science and engineering and the different grasps on the concept of technology, it involves important issues such as the status of the philosophy of technology, research traditions, and research methods. Research on technological artifacts has been a major concern for Chinese scholars in recent years. The "empirical turn" of philosophy of technology focuses on the analysis of artifacts of technology. It needs to go deep into the cultural context of specific technology and artifacts to analyze this relationship, so as appeared the return from the initially away from culture to the cultural study of technology. From the dimension of history and the perspective of axiology and cultural science, Chinese scholars carried out systematic research on the ontological level of technological artifacts in order to comprehensively interpret the dynamic relationship among the structure, meaning and existence of technological artifacts and reveal the influence of cultural factors in the process of design, formation and transfer of technological artifacts, including the ethical and moral factors involved in the design and use of artifacts.

The field of technology epistemology and methodology mainly involves the structure of technology cognition, the mode of technology development, the form of technology knowledge, the relationship between technology subject and object, the characteristics of technology thinking, the general process of technology invention, creation, development and application, the similarities and differences between technological method and scientific method, technological test and scientific experiment, problems such as technology forecast and evaluation, design and implementation of technological plans and implementation steps. Currently, the study of technological epistemology mainly focuses on five aspects: the definition of technological knowledge; the formation of technological knowledge; the nature of technological knowledge; the contextual nature of technological knowledge; the locality of technological knowledge. In terms of technological methodology, there are three main approaches, the first is from the focus on technological inventions to the study of technological design, the second is from the social and economic application of technology to the use of technology, and the third is technological evaluation throughout the process of technological design and use, including ecological orientation, the philosophical epistemological basis of the design method and the study of phenomenological implications.

The research topic of technological axiology mainly focuses on the characteristics of technological value, the duality of technological value and technological applications, social criticism of technology, technological alienation, technological optimism and technological pessimism, and Marxist technological values, etc. The specific background of the study of technological axiology concerns the relationships between technology and economy, technology and politics, technology and culture, technology and education, technology and ethics, technology and psychology, technology and policy, new technology revolution and contemporary society, etc.

As far as sustainable development and ecological technologies are concerned, the current academic research in China mainly includes energy and resource issues, environmental issues, ecological ethics, and related ecological technological issues. The discussion in this field is relatively abundant. It is also closely linked to the current status of science and technology development in China and the social issues. Moreover, in methodology, it shows multiple aspects such as technology economics, technological sociology, technology policy, and green technology innovation.

For technological innovation Chinese scholars mainly focus on the nature of technological innovation, the subject and object of technological innovation, the process and motive of technological innovation activities, the technological innovation mechanism, the value, the knowledge and the external conditions of technological innovation, and the method of technological innovation. Philosophical surveys of technological innovation also interact with the study of technological innovation from the perspectives of economics, sociology, management and policy, showing a pluralistic trend of technological innovation research.

On the comprehensive study of philosophy of technology, Chinese scholars pay more attention to the history of technological thought, the nature and status of philosophy of technology, the division of the system of philosophy of technology, and the relationship between philosophy of technology and related disciplines. This research has significant meaning for determining the future trend of Chinese philosophy of technology. However, compared the technological thinking with more social and practical significance, this research is not the mainstream.

In the comprehensive study of philosophy of technology, a more distinctive direction of Chinese scholars is social technology research. Social technology refers to a practical knowledge system created or invented (formulated) by a certain social government agency, accepted and used by all members of the society or a certain social stratum for adjusting the relationship between humans and society, solving social contradictions, and then realizing the purpose of safeguarding and reforming society. As an open system composed of personnel components and technology components, a social technology system includes physical elements, rational elements and procedural elements. The study of the form of social technology does not only help expand the research field of philosophy of technology, realizing the integration of the two academic traditions of philosophy of technology, but also promotes the new development of contemporary philosophy.

2. **Major philosophical issues in the development of contemporary technology**

Chinese scholars focus on the major philosophical issues in the development of contemporary technology in the following areas:

(1) **Philosophical issues in the development of nanotechnology**
In this field, Chinese scholars mainly focus on the risk of the use of nano-material and the deeper problems of the ecological environment and sustainable development, the ethical and social implications of nanotechnology from an ecological perspective, especially the impact on human survival and value realization; to realize the unity of a scientific spirit with a humanistic spirit by discussing "responsible innovation" and the technological design of nanotechnology informed by an ecological concept.
(2) **Philosophical issues in the development of biotechnology**
The main concerns of Chinese scholars in this field include the challenges to right to know and right to self-determination caused by the application of synthetic biotechnology in the field of medicine, and the timely adjustment of the ethics, regulatory, law and policy framework impacted by the philosophical reflection on biosynthetic technology. In addition, they also pay attention to the ethical issues brought about by the research and development of biopharmaceuticals, human trials and the construction of the human gene pool.
(3) **Philosophical issues in the development of information technology**
The main concerns of Chinese scholars in this field include the consideration of the new functions of information technology, i.e. the transformation of information technology from the intermediary of subject and object to the integration of subject and object, which leads to the redefinition of the relationship between subject and object, the source of human experience and the nature of cognition. The ethical problems of information technology are mainly manifested as "privacy" and "security". Chinese scholars focus on the ethical problems of technological design and the ways and means to resolve the dilemma. They also reflect on the ethical and philosophical problems of the "Internet of Things" in order to overcome the phenomena of advanced technology but a lag in ethics and policy, striving to provide a healthy and orderly environment for the development of information technology.
(4) **Philosophical issues in the development of cognitive science and technology**
The main concerns of Chinese scholars in this field include the relationship between artificial intelligence and human intelligence, ethical issues arising from the new advances in cognitive science and technology, risks and ethical issues in robot design and use, social and ethical issues arising from the enhancement of human brain function by cognitive technology, and so on.

(5) **Philosophical issues in the development of nuclear technology**
The main issues that Chinese scholars are concerned about in this field involve the technological risks, responsibilities and impartiality caused by the safety of nuclear power plants. It is necessary to reveal the characteristics of nuclear technology risks, the role of the media and public perception, and to find reasonable and effective countermeasures.

3. **Research on the philosophy of technology in Chinese context**

In the research field related to the philosophy of technology in a Chinese context, Chinese scholars mainly focus on the following areas:

(1) **Traditional cultural resources in Chinese philosophy of technology**

The study of the philosophy of technology in traditional Chinese cultural background is an important feature that distinguishes it from the study of Western philosophy of technology. Chinese scholars strive to analyze its historical role and its practical value, refine a series of issues and special categories that are unique to Chinese traditional philosophy of technology, and to enable research dialogues with contemporary Western philosophy of technology. Judging from the development of modern technologies, the traditional cultural resources in Chinese philosophy of technology are helpful in coordinating the contradictions among various elements in modern technological activities, developing and utilizing the tacit knowledge in modern technological activities, and forming the cultural characteristics of Chinese technological development, thereby improving the competitiveness of Chinese technology products.

(2) **The practical oriented paradigm in Chinese philosophy of technology**

The modern Chinese scholars of philosophy of technology focus on the practical oriented paradigm. The first generation of philosophers of technology represented by Chen Changshu not only had their own profound engineering professional knowledge foundation, but also advocated carrying out philosophy of technology research in the technological practice activities of enterprises, discovering and summarizing the philosophical problems at the forefront of technological development, carrying out in-depth ideological exchanges with engineers, and establishing practical oriented research paradigm. Changshu Chen's ideas of philosophy of technology is based on the practice of modern Chinese technology. His research on artificial nature studies, technological methodology, STS (Science, Technology and Society), Chinese national conditions, industrial philosophy, etc., all have distinct characteristics of being practice-oriented, and are worth exploring in depth.

(3) **Engineering practical oriented education of technology**

On the one hand, Chinese scholars try to integrate the research results of philosophy of technology into engineering education activities and cultivate the philosophical accomplishment, ethical awareness and social responsibility of engineering and technical personnel, on the other hand, they focus on the social and ethical problems brought about by engineering practice in China in the education of philosophy of technology, and cultivate the professional talents of philosophy of technology based on the engineering practice in China. Chinese scholars are currently working to integrate the research results of philosophy of technology into such courses as "Dialectics of Nature", "engineering ethics", "sociology of engineering" for science and engineering students, so as to improve the training quality of engineering professionals.

8.3 Strive to Create a School of Philosophy of Technology with Chinese Characteristics

1. **"Northeast School" is an important foundation for establishing a school for philosophy of technology with Chinese characteristics**

In the process of the development of philosophy of technology in China, Chinese scholars have been working hard to establish a school of philosophy of technology with Chinese characteristics, which is consistent with the general trend of the study of Chinese Dialectics of Nature (philosophy of science and technology) (Yu, 1996). In this process, some scholars in Northeastern China are called the "Northeast School" of philosophy of technology because they have the same research paradigm and long-term cooperation basis. In 2001, Professor Zeyuan Liu at Dalian University of Technology proposed "the Northeast School of Chinese philosophy of technology". He held that the philosophy of technology at Northeastern China represented by Professor Changshu Chen is a school that is quite unique in Chinese philosophical circle. It is rooted in the old industrial bases of Northeastern China, forming a community for philosophy of technology across different regions, building a unique theoretical framework and research paradigm of philosophy of technology, and creating a theory of philosophy of technological innovation with a distinct northeastern style (Liu, 2002) . We do not think that the "Northeast School" has actually formed, but if we regard it as the academic alliance of Northeastern philosophy of technology, the use of the term "Northeast School" in this sense is basically possible. Therefore, the "Northeast School of Chinese philosophy of technology" was jointly created by Changshu Chen and Deyu Yuan from Northeastern University, Shixu Guan from Harbin Institute of Technology, and Zeyuan Liu from Dalian University of Technology. As is widely known, Changshu Chen has made significant contributions to the development of Chinese philosophy of technology and the establishment and formation of the Northeast School of philosophy. As early as the 1950s, he created a theoretical precedent for the study of Chinese contemporary philosophy of technology, and published the "Pay Attention to the Methodological Problems in Technology" (Chen, 1957). Shixu Guan published an article in *Guangming Daily* in the 1960s to discuss the laws of the contradictory movement within machine tools which became an important beginning for the study of dialectics of engineering technology in China.

(1) **The 1980s is the germination period of the "Northeast school" of Chinese philosophy of technology**

In 1980, Changshu Chen proposed that the law of technology and technology development should be studied from a holistic perspective. Under the advocacy and leadership of Changshu Chen, the studies of Dialectics of Nature in Northeastern China mainly focused on the philosophical and methodological studies of technology. In 1982, he published the article "Difference and Unity of Science and Technology" in *Guangming Daily*. This was his first public article on the philosophy of technology. In 1983 he published "A Brief Discussion of Research in Philosophy of Technology" in the *Journal of Northeastern Institute of Technology*, which was the first article in China with the theme of "philosophy of technology". In 1985, he published "Technology is an Object Studied by Philosophy" in the *Journal of Dialectics of Nature*. He believed that philosophy of technology should become an independent branch of philosophy, opening the curtain for the study of Chinese philosophy of technology. In 1986, Changshu Chen published "The Impact of Technology on the Development of Philosophy" in the *Study for Dialectics of Nature*. He stated that fundamentally speaking, "Not the most universal philosophy or meta-philosophy has given birth to philosophy of technology, but rather the problems raised by the development of technology lead to the philosophy of technology. It is not that technological research needs a philosophical turn, but that philosophy research needs a technological turn" (Chen, 1986). From the aspects of ontology, epistemology and axiology, he discusses the influence of technology on philosophy and the evolution of the concept of technology. Especially in 1986, he jointly published *On Technology* with Deyu Yuan and organized the translation of *Introduction to Philosophy of Technology: Analytical Philosophy of Technology*, *Thinking Structure of Technological Science*, and other Western masterpieces. At the same time, in the middle 1980s, Changshu Chen conducted the engineering methodology study; Zeyuan Liu carried out the study of technology development methodology; Shixu Guan launched the Structure and Development Research on Engineering Technology. The systematic research on the philosophy of technology in the Dialectics of Nature atmosphere of Northeastern China has laid a solid theoretical foundation for the establishment of the Northeast School of Chinese philosophy of technology.

(2) **The 1990s was the formation period of the "Northeast School" of Chinese philosophy of technology**

Since 1993, led by Changshu Chen, Deyu Yuan, Shixu Guan and Zeyuan Liu, Northeastern University's doctoral degree authorization program, as the first doctoral degree program of philosophy of technology in China, takes the philosophy of technology as the main research direction, and has cultivated a large number of high-level theoretical talents in philosophy of technology. The graduates have become academic leaders in the direction of philosophy of technology in some universities and research institutes in China. Changshu Chen also edited the *Philosophy of Technology Ph.D. Series books of Northeastern University.* Up to now, it has published six series of nearly 60 books. In 1999, Changshu Chen published the first domestic monograph of philosophy of technology, *Introduction to Philosophy of Technology* (Chen, 1999, preface). Although he affirmed that this book "slightly approaches the engineering philosophy of technology, it is far from the humanistic philosophy of technology". However, Zeyuan Liu believed that the book regards artificial nature as the basic category of the natural transformation theory and the logical starting point of the philosophy of technology, both examining the technology from the perspective of the relationship between human and nature (the first half), and analyzing the technology from the perspective of the interpersonal social relationship (the second half). Therefore, it is a systemic synthesis of the philosophy of technology both engineering tradition and humanities tradition, as well as the dialectical unity of the relationship between human and nature and between human and human. It forms the paradigm of philosophy of technology of the overall inspection technology and reflects the style and characteristics of the Northeast School, so it can be regarded as an important symbol of the formation of the Northeast School of Chinese Technology Philosophy (Liu, 2002). In addition, Deyu Yuan and Shixu Guan devoted themselves to the practice of enterprise innovation in the old industrial base in Northeastern China, adhering to the paradigm of research on the philosophy of technology that is unified by the relationship between humans and nature and human and human. Research on technological innovation that combines both the engineering and social level has formed a "technological innovation philosophy". Moreover, it also reflects the unique style and characteristics of the Northeast School.

(3) **The early 21st century was the development period of the "Northeast school" of Chinese philosophy of technology**

In 2001, Changshu Chen published an article on "Maintaining the Vitality of Philosophy of Technology", emphasizing that "there is no status without discipline characteristics, no level without basic research, and no future without practical application" (Chen, 2001). Changshu Chen and Deyu Yuan also published the article entitled "On the Research Program of Philosophy of Technology" in 2001 to discuss the demarcation of the relationship between technology and science, the demarcation of philosophy of technology and philosophy of science, and the core issue that the study of philosophy of technology is technological epistemology or technological axiology (Chen & Yuan, 2001).

Changshu Chen also published the article entitled "Introduction to the American Philosophy of Technology" (1996), which systematically introduced the rapid development of American contemporary philosophy of technology over the past 30 years (1975–1995). He pointed out that "If you do not attach great importance to the development of European and American philosophy of technology and do not try to track the frontiers of the discipline, it is impossible to put forward the achievements on the basis of the predecessors and others. It is almost impossible to engage in the study of this discipline" (Chen, 2002). Zeyuan Liu published *The Study of German Philosophy of Technology* (2005) and conducted an in-depth study of the history and current status of the philosophy of technology in Germany, the birthplace of European philosophy of technology. Changshu Chen once provoked a profound thought, "If Chinese philosophy of technology starts from the early 1980s, it will take another 30 years or so. What will be the situation like?" Until now, half a century has passed since the publication of Professor Chen's first paper on philosophy of technology "Pay Attention to Methodological Problems in Technology" (1957). From the birth of the development of the philosophy of technology of the Northeast School, 40 years have passed. The study of philosophy of technology in China embodies the progress of the times and the transmission of context. In the 21st century, Chinese philosophy of technology should base itself on localization, develop towards globalization, and achieve sinicization, from the Northeast School to the philosophy of technology school with Chinese characteristics.

2. **Basic principles of creating a school of philosophy of technology with Chinese characteristics**

We believe that the following basic principles should be adhered to in order to create a school of philosophy of technology with Chinese characteristics.

(1) **The principles of combining understanding of development of emerging technologies with deepening understanding of traditional technology**

Technology is the object and foundation of the study of philosophy of technology. Human society is moving from an industrial society to an information society. The philosophers of technology at China should combine with our national conditions, such as the "rejuvenation of the northeast old industrial base" and other practical needs, deepening the understanding of traditional technology. At the same time, we must advance with the times to form a new delight in raising problems. At present, there is a worldwide focus on "NBIC technology" research. Western scholars have begun to examine the philosophical significance of the four major converging technologies. In order to establish a school of philosophy of technology with Chinese characteristics, it is necessary to understand the development of new technology and deepen the understanding of traditional technology, and to explore the internal relationship and transformation process between them.

(2) **The principle of combining understanding the "foreign philosophy of technology" with confronting contemporary Chinese practice**

As a branch of philosophy, philosophy of technology should be coordinated with the development of the whole philosophy at home and abroad, including understanding the progress of other branches of philosophy. However, the purpose of knowing western philosophy of technology is to produce research results with Chinese characteristics, instead of falling into the trap of "absence of Chinese characteristics" caused by "complete westernization tendency". This requires us to change from the "non-reflective followers" of the foreign philosophy of technology to the "reflective, autonomous thinkers". At the same time, we must move from disciplines-oriented philosophy of technology to problem-oriented philosophy of technology, and use the broad, interdisciplinary, comprehensive philosophy of technology to face the current practice of China in the midst of transformation.

(3) **The principle of combining the "empirical turn" with "experienced sublimation"**

The study of Chinese philosophy of technology based on the Chinese practice of technology, but it has some deficiencies, that is, abundant empirical researches with insufficient theoretical analysis. Therefore, it is very necessary to pay attention to the development of Western philosophy of technology and learn from the developed countries in the West. At the beginning of the 20th century, the study of European and American philosophy of technology was facing an "empirical turn". "Empirical turn" does not regard the philosophy of technology as an empirical undertaking, thereby losing its "philosophical" character, nor does it eliminate the normative and ethical values in philosophy of technology. Instead, it should turn from mainly focusing on technology use to the stage of design, development and production; less attention should be paid to abstract presuppositions and fictional myths of related technologies, while the analysis must be based on technical engineering practice; turning from theoretical normative to empirical descriptive, that is, to open the "black box" of technology, where the criticism and regulation of technology must be based on reliable experience. To establish a school of philosophy of technology with Chinese characteristics, on the one hand, we need to absorb the resources of "empirical turn" and integrate them organically with our own research tradition of emphasizing engineering practice; on the other hand, we should strengthen the theoretical analysis, realize the "experience sublimation", and combine the "physical" experience tradition with the "metaphysical" philosophical speculation.

3. **The path selection for creating school for philosophy of technology with Chinese characteristics**

(1) **Based on localization: the practical foundation of Chinese philosophy of technology**

In 1992, Changshu Chen proposed to pay attention to "STS Research and Chinese National Conditions". He pointed out, "the STS has its universality. It also has its particularity in different times and in different countries. Chinese STS research should pay attention to these two aspects, with particular attention to the current national conditions in China" (Chen, 1992). This is also applied to the study of philosophy of technology. The "Chinese national conditions" here refers not only to applying the general theory in the study of philosophy of technology to Chinese specific practice, but also to elicit new topics that need to be explored with regard to Chinese national conditions and gain new insights with practical significance. For example, the topic of "technological choices and leaping development in western ethnic regions", "technological innovation and the revitalization of old industrial bases in Northeast China", and so on, are realistic issues needing the attention of the localization of Chinese philosophy of technology. At present, China is implementing the "innovation-driven" national strategy. Therefore, we should base ourselves on localization, combine the theoretical innovation of philosophy of technology with the practice of "innovation-driven" construction in China, and promote the harmonious development of China's technology and society. These are the responsibilities of scholars engaging in research of philosophy of technology.

(2) **Internationalization oriented: a theoretical horizon of the school of philosophy of technology with Chinese characteristics**

The research of philosophy of technology in various eras and countries has its own international, regional or national perspectives. If the localization research concerning Chinese national conditions is to reflect the regional or national character of Chinese philosophy of technology, then academic exchanges between Chinese scholars and external scholars around the world are to reflect the international perspective of the study of Chinese philosophy of technology. Changshu Chen once said, "There is no status without discipline characteristics, no level without basic research, and no future without practical application". For current Chinese philosophy of technology, these are still applicable. At the same time, we have the saying: "There is no development without openness". The so-called opening up requires an internationalized theoretical perspective. The research on the basic theory of Chinese philosophy of technology must be based on localization and in line with international academia, drawing on foreign theoretical research results to promote theoretical innovation in the study of Chinese philosophy of technology and promote the dialogue and exchanges between Chinese and Western scholars.

(3) **Promoting sinicization: the future development of school of philosophy of technology with Chinese characteristics**

The sinicization of the study of philosophy of technology is based on Chinese scientific and technological practice. Under the guidance of Marxist philosophy, it learns and absorbs the theory of Western technology philosophies, and then looks for a link between philosophy of technology and Chinese national conditions, thus gradually establishing a theoretical system of philosophy of technology with Chinese characteristics and promoting the benign and sustainable development of Chinese technology. The sinicization of the study of philosophy of technology needs to understand historical background and the practical situation of Chinese development of technology. Globalization is both the generalization of technological logic and the specialization of different regions and cultures under this universal condition, which is a remarkable feature of contemporary technological development. This being so, how to adhere to the study of the philosophy of technology in the Chinese context is of great significance to a developing country with strong technology support.

The sinicization of philosophy of technology should also use Marxist theory, especially Chinese Marxist theory to guide the research of Chinese philosophy of technology. This is a distinctive feature of the study of Chinese philosophy of technology that differs from its counterpart in the West. To adhere to the Marxist tradition of philosophy of technology, first of all, we need to study the philosophy of technology thought of Karl Marx, Friedrich Engels and Chinese Marxists; then we need to insist on using Marxism to analyze and absorb the rational core of Western Marxist philosophy of technology, to guide the research of China's philosophy of technology.

The tradition of Western philosophy is based on the philosophy of mind-matter dualism which is concerned with the transcendental world, while the philosophical and cultural tradition of China is the human-oriented real world, and it focuses on the natural view of harmony between humans and nature. Chinese traditional philosophical culture not only contains the intention of seeking truth and knowledge, but it also has the orientation of seeking kindness. One of the fundamental characteristics of the natural view of Chinese traditional philosophical culture based on Confucianism and Daoism is the idea of advocating that the "*Dao* follows the law of its intrinsic nature" and "harmony between humans and nature". Of course, rich traditional Chinese philosophical resources are not directly easy to use, but require creative transformation. Therefore, this requires the study of contemporary Chinese philosophy of technology to have a profound grasp of the spirit of traditional Chinese philosophy and culture in respecting nature and seeking truth and kindness, thus truly "adapting ancient forms for the present day use".

(4) **Philosophy of technology focusing on Chinese issue**

The Sinicization of philosophy of technology needs to focus on Chinese issues, especially the values, ways of thinking and strategic policy orientations directly related to current technological development in China. This requires us not only to apply the general theory of philosophy of technology to specific Chinese practices, but also to elicit new topics that need to be explored in light of Chinese national conditions, so as to obtain new practical insights, and to provide important feasibility reports for government decisions. For example, issues such as the "fourth industrialization and China", "Made in China 2025", "big data", "cloud computing", and "intelligent robots" are all major practical issues that the research on Chinese philosophy of technology should respond to. The philosophy of technology that focuses on Chinese issue often has to face a wide variety of practical issues that are extremely comprehensive. Therefore, we should not only develop a professional philosophy of technology, but we also need to carry out interdisciplinary research.

8.4 Research on the Socialization of Technology: Theoretical Exploration of Chinese Philosophy of Technology

In the course of the study of the basic theory of contemporary Chinese philosophy of technology, we proposed the theory of technological socialization. This is a beneficial attempt to build an interactional model between technology and society in the process of Sinicization of philosophy of technology. In the monograph of *Introduction to Technology Socialization* (China Renmin University Press, 1995), Fan Chen systematically discussed the concept, essence and function of technology socialization, the social role of technology, the process of technological socialization, social integration and adaptation of technology and other basic concepts and theories. In the preface of the book, Professor Changshu Chen writes that this book "is of great significance to the formation and development of the sociology of technology" because it "constitutes a fairly complete system of technological socialization" and "can become a Chinese sociology of technology monograph" (Chen, 1995, I–III).

We believe that the essence of technological socialization is to make technology become accepted by society and to be identified by the public as a process that's socially compatible through the social integration of technology and the social adjustment of public psychology. Specifically speaking, on the one hand, under the integration of social location, the technology will meet and adapt to the requirements of social norms. On the other hand, through the adjustment of social psychology, the public has formed a positive social attitude toward technology, and finally made the technology accepted by society in the process of development. In this way, it is recognized by the public as a socially compatible technology, and this is the socialization of technology. The acquisition and improvement of technical and social attributes and the formation and development of technological and social roles constitute the basic content of technological socialization. Technological socialization is the process of obtaining and perfecting the attributes of technological societies, and is the process of the formation and realization of the technological role of societies (Chen, 1995, 5–8).

The realization mechanism of technological socialization lies in the social integration and social adjustment of technology. The social integration of technology not only refers to the process and state of maintaining an order among the various elements within the technological system; but more importantly, it refers to how the technological system as a result of social differentiation forms adaptation and coordination among heterogeneous elements of the external social system through certain ways and means: That it forming a socially compatible technology and realizing the process of technological socialization. Therefore, the process of social integration of technology should include many aspects such as social construction, selection, debugging and control of technology. Correspondingly, the view of social integration of technology should also include the basic theories of sociology of technology, such as social construction of technology, social selection of technology, social shaping of technology and social control of technology.

The social adaptation of technology mainly refers to the adjustment of social and technological psychology. In the process of technological development, people usually show a certain psychological orientation, that is, the formation of some social psychological processes and psychological phenomenon and behaviors related to technology, such as social perception of technology, social emotions, social motivation and social attitudes, etc. which is known as the technological psychology of society. In the process of technological socialization, the adjustment of social and technological psychology is mainly carried out through public media and legal norms. Appropriate socio-technological psychological adjustment not only enhances the public's awareness of science and technology, but also balances the public's emotional response to technology. Moreover, it will be very beneficial to form the social driving force for technological development and realize social acceptance and recognition of technology. The theory of socialization of technology is a reflection of the interactive relationship between technology and society on the basis of sociology, emphasizing the integration of natural, political, economic, and cultural and other local factors and technologies, and the social adaptation by means of public media and legal norms. "Philosophers only interpret the world in different ways, and the problem lies in changing the world." Combining the current state of technology development in China in the 21st century, the theory of technological socialization also analyzes the actual issues such as the socialization of the subject of technology in China, traditional technologies and their socialization, high technology and their socialization in China.

The theory of technological socialization is only a beneficial attempt for the sinicization of philosophy of technology (we should also study the socialization of science and the socialization of nature). But now the more important work is how to analyze the foreign philosophy of technology, and at the same time, to carry out research on the philosophy of technology that adheres to the tradition of Marxism, rooted in Chinese cultural traditions, based on the Chinese reality, focused on China's problem, and thereby gradually forming a theoretical system of philosophy of technology with Chinese characteristics, Chinese style, and in the Chinese context.

8.5 Conclusion

The path of choice for philosophy of technology in China must first be based on localization. This is the practical foundation for the sinicization of philosophy of technology. Secondly, it must be internationalization-oriented, which is the theoretical perspective of the sinicization of philosophy of technology. Thirdly, it shall gradually realize sinicization, which is the future direction of sinicization of philosophy of technology. As long as we conscientiously face the background of science and technology globalization, based on China's scientific and technological practice, achieve technological innovation of enterprises, enhance the ability of independent innovation, and serve the construction of innovative countries, the study of Chinese philosophy of technology will gradually form an academic community of philosophy of technology with a basic research paradigm, a clear research program, and scientific research methods. We must continue to adhere to the principles of "highlighting position by characteristics, reflecting the level by research, seeking the future by applications and promoting development by openness". The school of philosophy of technology with Chinese Characteristics will gradually occupy a place in the academic circle of international philosophy of technology.

References

Chen, C. (1957). Pay attention to the methodological problems in technology. *Journal of Research and Communication of Dialectics of Nature, 1*, 24.

Chen, C. (1986). The influence of technology on the development of philosophy. *Studies in Dialectics of Nature, 2*(6), 1–8.

Chen, C. (1992). STS research and China's national conditions. *Studies in Dialectics of Nature, 8* (1), 86.

Chen, C. (1999). *Introduction to the philosophy of technology* (p. 44). Beijing: Science Press.

Chen, C. (2001). Keeping the vitality of the research on the philosophy of technology. *Science, Technology and Dialectics, 18*(3), 43.

Chen, C. (2002). A brief introduction to the literature of philosophy of technology in the United States. In *Chen Changshu's collected works of philosophy of technology* (p. 97). Shenyang: Northeastern University Press.

Chen, C., & Yuan, D. (2001). Rethinking of the research program of philosophy of technology-discussion with Professor Huaxia Zhang and Professor Zhilin Zhang. *Studies of Dialectics of Nature, 17*(7), 41.

Chen, F. (1995). *Introduction to technology socialization*. China Renmin University Press.

Chen, F., & Chen, J. (2009). Retrospect and expectation on contemporary philosophy of technology in China. *Studies in Dialectics of Nature, 25*(10), 56–62.

Chen, W. (2005). Review and prospect of research on philosophy of technology in New China. *Journal of Northeastern University (social science edition)*, 7(4), 235, 236–237.

CPC General Branch Committee of Machine Tool and Automation, Department of Mechanical Engineering, Harbin Institute of Technology. (1960). Looking at the contradiction movement law of machine tool from the "Block type machine tool". *The Red Flag, 1*(24), 27–30.

Gong, Y. (1991). History of dialectics of nature (II). *Studies of Dialectics of Nature, 7*(2), 4.

Liu, Z. (2002). On the northeastern school of philosophy of technology in China. *A yearbook on the study of Chinese philosophy of technology in 2001* (pp. 134–141). Dalian: Dalian University of Technology Press.

Qian, X. (1957). Methodological problems in technical science. *Journal of Research and Communication of Dialectics of Nature, 1,* 34.

Yu, G. (1996). *A school of philosophy is rising in China*. Nanchang: Jiangxi Science and Technology Press.

Chapter 9
Reflection on the *Dao* of Technology: Philosophy of Technology from the Perspective of Chinese Culture

Qian Wang and Wei Zhang

Abstract Reflecting on the *Dao* of technology, as a kind of philosophy of technology from the perspective of Chinese culture, can reveal the secluded characters in terms of the essence and evolution of technology. The development of Western modern technology goes through four stages successively, includes science-based, artistic design, orient-ecology and moralization. It verifies the theoretical values of the *Dao* of technology indirectly, which provides a new approach to understand the relations among related elements in modern technological activities, and inspires a comprehensive development of technology in future.

Keywords Chinese culture · *Dao* · Organism · Holism · Philosophy of technology

9.1 Introduction

Philosophy of technology has seen various general analyses of technology. Some scholars have focused on the nature of technology through logical analysis, and others have employed a more organic approach. This article argues that those researchers to some extent did not adopt a holistic and organic view when analyzing technology and the connections to its related elements. While the concept of *Dao* which from the perspective of Chinese philosophy can do it.

Some scholars of the former Soviet Union once interpreted the nature of technology from the aspect of means of labor and equipment, believing "technology to be means of labor, instrument of production and everything substantial used to

Q. Wang
Department of Philosophy, Dalian University of Technology, No. 2, Linggong Road, Ganjingzi District, Dalian, China
e-mail: qianwang@dlut.edu.cn

W. Zhang (✉)
School of Marxism, Central China Normal University, No. 152, Luoyu Road, Wuhan, China
e-mail: hnhszw@163.com

Q. Wang (ed.), *Chinese Philosophy of Technology*, Philosophy of Engineering and Technology 34, https://doi.org/10.1007/978-981-15-1952-9_9

improve the efficiency of labor". Some European scholars interpreted the nature of technology from the aspect of knowledge and skills. For example, Yves R. Simon regards technology as the knowledge on how to act and how to achieve the goals of human beings. Simon M. Bunge (1951) sees technology as "the totality of knowledge used to control the process of hanging and creating things of nature or of society, which is driven by some valuable practice and conditioned by scientific methodology" (Bunge, 1967). On the other hand, humanists are used to understanding the nature of technology from the relationship between technology and society. For instance, Heidegger interprets technology as a sort of *Un-verborgenheit* (un-concealing), and hereby modern technology urges to reshape things away from their original nature, namely *Ge-stell* (Heidegger, 1977). Andrew Feenberg and Frankfurt School members emphasize its ideological role and political feature; Erich Fromm pays attention to the control of technological civilization over individual emotion and rationality. In *Thinking through Technology*, Carl Mitcham summarized the views on its essence in four forms: technology as object, technology as knowledge, technology as activity and technology as volition (Mitcham, 1994).

Also, some researchers adopt the perspective of organism philosophy to discuss the nature and evolution of technology in the history. Aristotle took technique as the extension of the functions of hands. As he put in his *History of Animals,* hands are more of a multifunctional means for using other tools than serve as mere a tool. And based on that, Ernst Kapp proposes the famous "Organ Projection Theory". Karl Marx also proposes a similar theory known as "Organ Extension Theory", which implicates that technology, as the extension of human organs, is the externalized embodiment of human's character. Arnold Gehlen believes that technologies are not only a supplement to hands, but also a reinforcement of them. Donna Haraway influenced by Whitehead's philosophy, analyzed the characters of "Cyborg" as the man-machine interface, indicating that it bears the properties of both human and technology and realizes their co-evolution (Haraway, 1991). Nonetheless, all these explorations, to some extent, deal with the organism philosophy from the perspective of instruments and machinery alone without a holistic view on the organic relationships among operators, instruments, objects and environment in the technological system.

The inspiration from Chinese cultures may help to open a new approach to make up for this defect. *Dao* in Chinese cultures is a critical tool to achieve such a holistic and organic view. *Dao* is very important in Chinese philosophy. Some scholars consider it similar to the elements composing the world—"water", "fire", "earth", "air" in ancient Greek natural philosophy as the origin of the world, or the category of "logos" as law and order of the world. But these comprehensions fail to catch the true meaning of *Dao* in Chinese cultural context. *Dao* can be indicative of the nature of substantive human practices and most instructive in rationality and optimization of technical activities. The relation between *Dao* and *Ji* (技, technique) not only has profoundly influenced the development of Chinese traditional technology, but also bears important inspirations to explain the evolution of Western modern technology.

The awareness of Chinese culture on the relation between *Dao* and *Ji* shows significant value on revealing the organic connections of varieties of elements of technology. Technological practice of human being is a complicate system which consists of varieties of elements; any single element cannot fully represent the essential character of technology. The guidance functions of *Dao* for development of technology emphasis the harmony of related elements in technical activities and the progress of technical progress on the whole. Without the guidance of *Dao*, various technical elements would be in disharmony with each other, and humans will use any method for the sake of utilitarian purpose, and this may result in the alienation of technology.

This article is to first introduce Chinese thinking in terms of *Dao* in Sect. 2. In Sect. 3 we show the benefit of considering the development of Chinese technology with *Dao*, addressing in part the Needham problem. In Sect. 4 we move to modern technology. We describe its development and show that understanding this development in terms of *Dao* has advantages. In the conclusion section, we show the benefit for existing philosophy of technology to introduce that concept of *Dao*.

9.2 *Dao* and the Essence of Technology

Dao is associated with technical activities in its original sense, *that is*, "road". According to textual research, one part of *Dao*'s Chinese pictographic character "道" is "首" (head), and the other part is "走" (walk) (Needham, 1956). *Dao* shows the function of roads, namely, man could walk from here to there with the guidance of the "head". This process includes the purpose, direction and procedure. Other human practical activities also have the similar features, and they adopt such terms as "by means of" and "steps" to do the description. This feature could also be found in English. Specifically, the English word "way" carries the meaning of both road and means or method.

However, *Dao* as a philosophical category refers to neither walking on the road, nor various specific means or methods in human practices. In ancient China, the specific means or methods is named *Ji* (技, technique) or Shu (术, Skills), instead of *Dao*. *Dao* is highly generalized to the essential characteristics of the means and methods of human practice. In term of one technical activity, this nature is that all kinds of technical activities consist of a series of coherent steps or links, where different operators are required to respectively use specific means and methods and therefore, they may arrive at different results. Yet, there must be one optimal and most reasonable mean or method that follows the essences of this kind of activity, which is called *Dao*. And the so called "most reasonable" means technical practice should be in accord with the objective feature of the relationship among related elements of technical activities, which is embodied in the harmony among various technical elements. And the so called "most optimal" means less consumption of labor, high quality and high efficiency oriented towards social value and interests, this is the "the craftsmen should grasp artistry" recorded by *Kao Gong Ji* (考工记),

a Chinese ancient book on techniques. The most reasonable means and method determines the optimal result, which in turn, seeking the optimal result constantly promotes more appearances of new reasonable means or methods.

Dao is an objective existence, which exists in the interactivity between such technical elements as operators, tools and objects. It is the ideal state of techniques and transcends various specific techniques. In the Chinese Daoist classic, *Chuang Tzu*, it illustrates the ideal state of technique with such a story:

A man who was a cook for Prince Wen Hui (文惠) was cutting up a bullock. The blow by his hand, the thrust from his shoulder, the stamp of his foot, the heave with his knee, the whish of the flesh coming away, and the whistle of his knife going in were all perfectly in time, having the rhythm of the Dance of the Mulberry Grove, and exactly in time like the chords of the Ching Shou (经首). The Prince said, "How excellent, that you should reach this pitch of perfection!" The cook put down his chopper and said, "The thing that your humble servant delights in is something higher than art, namely the *Dao*" (Hughes, 1942). In his response, the cook points out the gist of the relationship between *Dao* and *Ji*. Going after the *Dao* above *Ji* implies to elevate various means and methods to the ideal state which is in accordance with the natural essence of things. Once *Dao* is achieved, men could use tools skillfully and technical operations would follow the natural trend, which is what Lao Tzu calls "*Dao* follows nature".

Compared with the concrete means and method in technical activities, *Dao* is highly abstract. But man could still approach *Dao* perpetually and interpret its power and values when committing to harmony with nature. Lao Zi thinks that there is no way of finding where *Dao* starts from and where it ends up. Guan Zi suggests that *Dao* is something that we cannot see, we cannot hear; yet there is a course to its accomplishments. In modern terminology, *Dao* is an optimal "program" of evolution or generation of things, which is in harmony with their natural essences.

To consider *Dao* as the natural, the most reasonable and optimal way is not confined to the concrete operation processes or technical indication of less consumption of labor, high quality and efficiency. In the view of ancient Chinese philosophy, *Dao* in the concrete technical operations is just "small *Dao*" (小道) or "byways". Confucians believed that "although some byways can lead a person to someplace desirable, it fails to direct a man aiming at the destination afar. Therefore, the gentleman does not take the byways." On one hand, it affirms the value of byways; on the other hand, it also indicates its limitations. To attain reasonableness and optimum, *Dao* requires extending by integrating all kinds of social and cultural elements related to the technical activities. Such is what is called the grand *Dao* (大道).

The existence of *Dao* of technology is based on the organic connections of the related elements in technological system. Humans as the subject of technical activities, have the organic characters in terms of physiology, society, and spirit. They also endow the nature with these organic characters in the process of technological operations, make them humanized and influence the natural and social environment, causing varieties of conflicts which need to be resolved in a coordinated way, so that the development of technology needs to be conducted by *Dao*. To comprehend the guidance of *Dao* for *Ji* needs a holistic and intuitional way of thinking in which the organic connections between kinds of elements in technical activities could not be understood in a comprehensive way by logical analysis. The analytical way of thinking guides people to pay attention to the local "present" relations, but ignore the "absent" integral and secluded relations. The guidance of *Dao* for *Ji* helps people realize the existence of these relations and adjust their behaviors reasonably. The harmonious relationship among elements in technical activities which is revealed by *Dao* of technology needs to be understood by a holistic and intuitional way of thinking.

Generally, the harmony of all elements relevant to technical activities includes the following aspects:

(1) The harmony of technical operators and tools. Notably, tools here are not invented as an extension, projection or supplement of human organs; instead, they are designed to conform to the physiological and psychological features. When the craftsmen operate their tools skillfully, the tools work like an organic part of their bodies. In ancient China, much attention was given to the coordination between the physical functions of tools and the physiological characteristics of the operators in the design. According to *Kao Gong Ji*, for instance, when designing a carriage, one factor that should be firstly taken into consideration is whether the diameter of the wheels is convenient for men to get on and off the carriage, and after this is to determine other things, e.g. the best proportion of the wheels to other parts, are then considered (Lin, 1990). In *Dream Pool Essays* (梦溪笔谈), the author mentions that in designing the size of steps in ancient palace, the craftsmen should consider whether it is convenient for the carriers of the sedan of emperor to walk on and they figure out different forms of steps—steep-gradient steps, easy-gradient steps, and flat steps (Shen, 1998). The modern ergonomics emphasizes the coordination of the working situation and human's physiological and psychological features. This idea has already been embodied in ancient Chinese technical practices in a holistic and intuitional way.

(2) The harmony of body and mind of operators. Many outstanding craftsmen in ancient China considered technical activities as an art. Thus, the experience, skill and knack of the craftsmen are called "art of technique", "art of hand" and "consummate art" in ancient China. They believed that technical activities were pleasant to their body and mind as they involve elegant acts, saving labor, high efficiency and exquisite products with constantly high quality improvement.

(3) The harmony of people in technical applications. *Dao* in technical practice bears moral implications. It requires the operator to find out the most reasonable and optimal method when dealing with the relations between human and things, between man and man. They should bear hardships and stand hard work, being

strict and earnest, cherishing the kindheartedness, and bring benefits to the common people. It demands the cooperators in technical practices to be honest and harmonious. Only by achieving those requirements, could the craftsmen be called the noble with *Dao*.

(4) The harmony of technical activities and social life. In developing technology, ancient Chinese emphasize the principle of "techniques should serve for statecraft and pragmatism". Confucians do not mean a firm refusal to develop techniques but object to merely indulging in diabolic tricks and wicked crafts. They advocate regulating technique by *Dao* so as to harmonize the development of technology with other social elements for the purpose of making the social life stable and safe.

(5) The harmony of practical activities and natural environment. *Dao*'s guidance on technique aims to make technical activities follow nature and enable men's activities to be of help in maintaining the dynamic stability of nature rather than destroying it. Therefore, it should try to exert the role of natural factors in the process of technical operations, and make full use of the renewable resources and energy (Wang, 2005).

Only in the harmony with all those respects above, can technological practices meet the requirement of "grand *Dao*". If we consider the relations between operators and tools or relations between man and man in technical applications alone, this is just a partial reasonableness and optimum, *that is*, "small *Dao*". There is still a long way to go before reaching holistic reasonableness and optimum. And in that case, it is hard to avoid the ignorance of social benefits and environmental disruptions.

Thinking from the relation between *Dao* and *Ji* captures the essential feature of technical activities comes from a special perspective. Previous scholars once interpreted the essential features of technical activities from different perspectives. However, whether those scholars understood the nature of technology from means and equipment with a belief that technology is a means of labor, a tool of production and everything substantial used to raise the efficiency of labor, or from its function emphasize its role in remaking nature, the organic connections among these technological factors and tendency of technological development are not considered fully. Now, the value of this work is to integrate all of these into a holistic view.

The awareness of Chinese cultures on the relation between *Dao* and *Ji* reviews a high value on revealing the organic connections of varieties of elements of technology. Technological practice of human beings is a complicate system which consists of varieties of elements; any single element cannot fully represent the essential character of technology. The guidance functions of *Dao* for development of technology emphasizes the harmony of related elements in technical activities and the progress of technical progress as a whole. Without the guidance of *Dao*, various technical elements would be in disharmony with each other, and humans will use any method for the utilitarian purpose, this may result in the alienation of technology.

9.3 *Dao* and the Development of Traditional Chinese Technology

The significance of the *Dao* of technology can be confirmed in two aspects. One is the direct confirmation, *that is*, the emphasis on the guidance of *Dao* for *Ji* which obviously promotes the development of traditional Chinese technology. The other is indirect confirmation, *that is*, the history of Western modern technology which verified the rationality and inevitability of *Dao* of technology.

According to the research of J. Needham, traditional Chinese technology stayed at a level which the western countries were far behind to catch up with. He said:

"Why should the science of China have remained, broadly speaking, on a level continuously empirical, and restricted to theories of primitive or medieval type? How, if this was so, did the Chinese succeed in forestalling in many important matters the scientific and technical discoveries of the dramatis personae of the celebrated 'Greek miracle', in keeping pace with the Arabs (who had all the treasures of the ancient western world at their disposal), and in maintaining, from the 3rd to the 13th centuries, a level of scientific knowledge not yet achieved in the West? How could it have been that the weakness of China in theory and geometrical systematization did not prevent the emergence of technological discoveries and inventions often far in advance (as we shall have little difficulty in showing) of contemporary Europe, especially up to the 15th century?" (Needham, 1954).

Why could ancient China receive a high level without the support of scientific theory? One important factor which deserves attention is the guiding function of *Dao* for *Ji*. This guiding function is not unfolded by the way from principle to examples, but by a holistic and intuitional culture atmosphere, using the methodological model of *Yin-Yang* (阴阳) and five elements, influencing the treatments on the relations among related elements in technical activities.

In fact, in ancient Chinese technical classics, the vast majority of the contents are about the concrete procedures and technical standards of technique and engineering activities, whereas only a small number of them are about the ideology of ancient technical officials and craftsmen. There are only a few of exceptions, for example, Sixie Jia (贾思勰), in the Northern Wei Dynasty, said in his *Qi Min Yao Shu* (齐民要术) (*Important Techniques for the People*): "if you follow the climate and consider the geographical conditions, you will get more benefits with less labor; If you follow your own inclination and violate *Dao*, you will otherwise get nothing even you do lots of work." (Jia, 1958)

Ancient Chinese technical inventions mainly adopted intuitive approach,lacking in the instructions of strict logical and methodological system in modern research, but due to such a goal as *Dao*, the deepening of intuition obtained its direction and became a process of pursuing *Dao*. Lao Zi emphasized the differences between "you wei" (有为, do) and "wu wei" (无为, not do). The former refers to acting by following the means and methods regulated by men, while the latter refers to acting by following the means and methods in accordance with the nature of things, namely accepting the guidance of *Dao*. All means and methods recorded in ancient

Chinese technical classics are the characteristics of the stage of "you wei". Once men could use them very skillfully, they could be mastered by mind instead of memory. *Dao* is the reflection of this stage of extreme skillfulness, just like "wu wei". So Lao Tzu says:

Learning consists in adding to one's stock day by day,
The practice of *Dao* consists in subtracting day by day.
Subtracting and subtracting yet, up to reach inaction. (Lao Tzu, 1999)

The development from *Ji* to *Dao* (由技至道) is a process of fully unfolding the physical and intellectual potential of human beings. Every technical activity needs to make use of some aspects of men's physical and spiritual abilities, at the same time, the experience, habits and the regulations on technical standards also confine or conceal some other aspects of those abilities. The process of pursuing *Dao* is one to keep on eliminating unreasonable confinement and developing men's physical and spiritual potential. In this process, the intuition played an important role. The duration and degree of heating, for example, is important for ceramics; the temperature of water plays a key role in filature; the matching effect of each part is significant in making carriages; all of them should meet the requirements of 'neither going too far nor too near', and all of these things mainly depend on human's intuition. The cognitive mechanism of the relationship between technique and *Dao* helps people obtain those technical tips fast, and this may be the key factor in explaining the highly advanced technique of ancient Chinese at the level of intuition.

Further, the theory of *Yin-Yang* and five basic elements offer a conceptual model for the intuitive thinking of traditional Chinese technical activities. Traditional Chinese medical science, agriculture and engineering technique and other things are all based ideologically on the theory of *Yin-Yang* and five basic elements, which present a particular feature of Chinese cultures. This theory per se could be explained literally, but cannot be equal to *Dao*. It serves as a spiritual tool to reveal and grasp *Dao*. Five basic elements refer to five materials (metal, wood, water, fire, earth). And this theory does not show the substantial relationship, but the functional relationship between these five elements. It provides rules and models for designing and fixing the specific ways, steps and processes of technical activities so as to enable them to keep on approaching the ideal realm of *Dao*. The views of mutual infiltration and supplement between *Yin* and *Yang*, as well as with identity of opposites, can be used to prevent technical development from going astray. This idea not only has advanced the development of medicine, health maintenance and architecture, but also is embodied in the inventive principles of compass, transmitting bronze mirror and movable-type printing (Xie, 1989). The mutual promotion and restraint between the five basic elements is helpful in building up a relationship of check and balance between elements related to technical activities. Influenced by the theory of *Yin-Yang* and five basic elements, many ancient Chinese inventions and creations are filled with changes and intelligence.

The guidance of *Dao* for *Ji* played an obvious role in coordinating ancient Chinese technique and natural environment. The idea of "*Dao* follows Nature" enabled Chinese traditional agriculture to focus on favorable climatic, geographical and human conditions. Especially the adoption of the method of "cultivating the earth" by using organic manure to improve soil fertility avoided the soil infertility after long farming and stabilized a high yield of grain and ecological environment. In the domain of traditional manufacture, such as spinning, food production, ceramics and papermaking, the main resources were natural materials, so there were few environmental pollution problems. Even the mining and smelting activities take natural factors into full consideration. For instance, when exploiting coal mine, ventilation, dewatering, timbering, backfill and the recycle of ferric sulfide are given special attention. The craft of "wrought steel" took loess and straw as slags. The carburizers of carburized steel were used to deal with pine and fermented soya beans. The guidance of *Dao* for *Ji* could also help coordinate the relation between technique development and social life. The cultural tradition of "inhibition of luxury" is to control the production and sale of excessive luxury artifacts with the aim of avoiding the social instability caused by such technical activities. Confucians constantly condemned the engineering which would waste manpower and money, considering the massive construction projects and pleasure-seeking of emperors as "Wu *Dao*" (无道) (going against the *Dao*). They believed advocating plainness to be the exact "Zheng *Dao*" (正道) (going with *Dao*) to administer state affairs well and ensure national security. In the situation of natural economy, the idea of "Yi *Dao* Yu Shu" (以道驭术) (to control technique by *Dao*) helped keep the harmony between Chinese traditional technology and social life as a whole.

9.4 The Development of Modern Technology from the Perspective of the Relation between *Dao* and *Ji*

The trend to *Dao* can be found through self-examination and adjustment. It can be taken as the indirect confirmation of the rationality and inevitability of *Dao* of technology. Seen from the relation between *Dao* and *Ji,* the development of the Western modern technology, takes on a trend of approaching the technological *Dao* in a self-reflective and self-adjusting manner. This can be taken as the indirect confirmation of the rationality and inevitability of *Dao* of technology. Broadly speaking, modern technology has advanced respectively on the scientific, artistic, ecology-oriented and moral basis. It is a process from pursuing the efficiency alone to pursuing the harmonious development of technology with nature, society and humans. Thus, it can also be seen as a parallel process of approaching "from *Ji* to *Dao*". However, this parallel development process charged a high price, if it was guided from *Dao*, the price would be largely decreased.

(1) Advancing on Scientific Basis: Embodiment of Optimization

With the advent of modern science in the West, the research outcomes of science begin to be applied into technical innovations, promoting the science-based technology. Essentially, the science-based technology focuses on the improvement in efficiency. It is the expression of instrumental rationality and the outcome of the modern economic system. Modern technology brings us enormous material well-being, but it sacrifices the benefits of goodness and beauty contained in ancient technique, leaving only the instrumental rationality. Confronting modern technology, human beings become "one dimensional man" (Marcuse, 1964), while nature becomes "a source of disposable raw material". When the mid-twentieth century, humans suddenly recognized that the spring supposed to be vibrant became silent; our living environment had become so precarious. "Only later was the potentially fatal flaw in the scientific foundation of the new technology discovered. It was like a two-leg stool: well founded in physics and chemistry, but flawed by a missing third leg—the biology of the environment." (Commoner, 1971) Modern technology not only missed the leg of ecology, but also missed the legs of art and morality. Therefore, the early modern technology had serious defects in its internal structure. Its value orientation should be relocated. The development of modern technology which is based on science pursues the optimization of economy, but it leads to many negative effects in other aspects, this is obviously the deviation from the *Dao* of technology.

(2) Advancing on Artistic Basis: The Harmony of Technology and Operator's Body and Mind

In order to adapt to the characteristics of mass production, product design must be standardized. This results in the monotony of products, losing its personality and aesthetic taste. Meanwhile, in order to satisfy the pursuit of factory owner for efficiency, designers usually neglect the traits of the body and mind of the workers, which results in great harm to the physical and mental health of workers. Repetition of monotonous movements and high intensity of labor even brought them serious occupational diseases. This situation was quite serious from the 19th century to the early 20th century. In this context, the English designer, William Morris launched the "Arts and Crafts Movement" in the 1880s with the aim of remedying the flaws of modern technology by art. With the influence of Arts and Crafts Movement, another art movement, "Art Nouveau", emerged in Europe in the 1890s. The genuine integration of technology and art came into being in Bauhaus, a school which was operated in German from 1919 to 1933. It generated the design style of "Modernism" which adopted the modern mass production. One of the main objectives of the Bauhaus was to unify art, craft, and technology. The machine was considered a positive element, and therefore industrial and product design were important components. The design concepts it proposed include: the interaction of technology and art; the purpose of design is for humans not for products; design must conform to law of nature (Forgács, 1995). With the advance of artistic design, modern technology rediscovered the dimension of beauty which was discarded by instrumental rationality in early modern times. This process shows an unselfconscious return to the *Dao* of technology after a series of faults.

(3) Advancing on Ecology-oriented Basis: The Harmony of Technology and Ecological Environment

"Ecology-oriented design" is to embed the ecological laws into technical design, making it compatible with natural environment and trying to reduce the influences on the natural environment as much as possible. In the mid-20th century, the awareness on environment protection was roused, the consequences of which were a series of global environmental movements. Sim Van der Ryn and Stuart Cowan claimed that the environmental crisis is, in some sense, caused by design. So it is crucially important to solve the environmental problem through design. Ecological design refers to "any form of design that minimizes environmentally destructive impacts by integrating itself with the course of life" (Van der Ryn and Cowan, 1996). The core idea of ecological design can be simply described as three "R": Reduce, Reuse and Recycle. Specially, we should follow the principle that design should be consistent with ecological principle and for site-specific context (Scott et al., 2001). This process reflects the further return of modern technologies to the *Dao* of technology. The harmonious idea between modern technology and natural environment is the result of thorough reflections on the accumulated ecological crisis.

The concept of ecology-oriented design has indeed greatly improved ecological environment. However, it has its limitation, which is to employ technological innovations to achieve sustainable development. Technology is actually just one of the contributing factors, and the way we use a technology also carries much weight on environment. An environmentally-friendly product, good intention as its designer is, is equally likely to be detrimental to environment if used inappropriately. To make a design function better in practice, it must consider both human and technological factors. That means that a design should not only be environmentally friendly, but be made after consideration about potential influences on its users.

(4) Advancing on Moral Basis: The Harmony of Technology and Society

The main problem of ecological design may be that it pays too much attention to the environmental effects of the products themselves, but seldom cares the possible impacts on the behavior of the users (Verbeek and Slob, 2006). For example, the invention of vending machines of low energy consumption leads to their proliferation in more locations; more efficient air travel has generated 'low cost' companies and therefore increased the number of passengers. Designers should not only focus on minimizing the environmental impacts of the products themselves, but also pay attention to what reactions the users will have on their products. So, it still needs "moralization". The moralization of technology has aroused much attention in recent years. It can be interpreted in two ways. From the perspective of externalism, it is to use morality to regulate the practice of technology, making the technical design an application consistent with the moral principles. From the perspective of internalism, it is to embody morality into the technological design, with the aim of conducting or regulating the behaviors of human beings when they are using these products (Verbeek, 2006). The designers can intentionally design artifacts to positively influence humans' behaviors. If ethics is about "how to live", then, designers actually engage ethics by other ways—"doing" ethics, they provide a "material answer" for this question (Verbeek, 2011). The moralization of technology enables technology to keep a harmonious relationship between human beings during the using process, and respects and protects the benefits of all kinds of stockholders. This process highlights the role of morality in modern technological activities, which is closed to the holistic requirements of the *Dao* of technology.

From the chronological point of view, stages of development on scientific basis, artistic basis, ecology-oriented basis and moral basis appeared one after another, the driving force for the transformations among these stages is the pressure to solve the prominent problems in the technological practices. In other words, they are the results after paying a certain price. However, the general trend of these stages is to realize the harmony among all kinds of elements in the technological system, which is in line with the principle of the relation between *Dao* and *Ji*. In this sense, it can be said that the rationality and inevitability of the *Dao* of technology are indirectly confirmed by the development of modern technology.

The development of modern technology is very hard to avoid the destiny from constant deviations to continuous regressions. When dealing with the related elements in technological practice, modern technology shows obvious tendency of calculation, compulsivity and importunity, on which Heidegger has contributed an in-depth analysis. This way of thinking results in a model of ceaselessly conquering and expanding. However, the bearing capacity of the earth is limited. And this limitation has yet to be obvious at the time of the rise of modern technology. With the advent of environment degradation, climate change and resource exhaustion, this limitation became much more obvious. In this situation, to reflect technology from the perspective of the relation between *Dao* and *Ji* is not a dispensable issue, but a necessary one. Nevertheless, it should be noted that the relation between *Dao* and *Ji* in Chinese tradition techniques has its defect too. They are limited in terms of precision, rigorousness and standardization and fail to translate into modern technology. Reflecting on the relation between *Dao* and *Ji* against the background of modern technology, it is obliged to eliminate these historical limitations and uncover the diachronic rationality and modern values of *Dao*. The expression on the relation between *Dao* and *Ji* should take the form which is easy for people to understand. The perspective of organism philosophy combined with the modern interpretation on Chinese philosophical categories can help people adequately understand the profound connotation of the *Dao* of technology, enabling this important intellectual resource to play a role in promoting the benign development of technology.

Comprehension on the relation between *Dao* and *Ji*, as a kind of organism philosophy of technology against Chinese culture, can play a complementary role with other schools of philosophy of technology. Because the way of thinking in modern technology is characterized by logical analysis, which takes Logos as an initiatory thought, this is also manifested in the complementary interaction between *Dao* and Logos.

The complementation of *Dao* and Logos mainly embodies in the fields in which Logos alone has not come to play. In other words, it embodies in the fields in which the development pattern of modern technology is defective. For example, in the era of environmental crisis, humans should keep the harmony among the elements of technological activates as a whole; it should avoid the mistakes which have been made in the process of modernization by developed countries in developing countries; the physiological and intellectual potentialities should be exerted fully to improve the technologically innovative ability. In these respects, we all need the guidance of *Dao* on *Ji*. Although the *Dao* of technology comes from China, its influence can be extended to the whole world.

9.5 Conclusion

According to the analysis above, it can be found that *Dao of technology* carries on important theoretical and practical significance in understanding the essential characteristics and the relations among elements of technological activities. The great achievement of ancient Chinese technologies, which is derived from intuitive experiences, is a clear manifestation of the value of *Dao* of technology. Yet, the problems in the development of Western modern technologies that is lack of holistic and organic idea also serve as a foil to the value of *Dao* of technology.

With respect to existing theories of philosophy of technology, what new inspirations will the idea of *Dao* of technology bring? And what approaches and methods should we take to implement and apply the idea of *Dao* of technology? These questions need further explorations. But, philosophy of technology with the core of the relation between *Dao* and *Ji* emphasizes the integrality and organic property, and tries to open up a new approach for the development of philosophy of technology. That is to play the role of intuitive experience in coordinating the relations among related elements in technological activities. Most of the existing philosophy of technology is based on the method of logical analysis. Although phenomenology and hermeneutics have referred to the role of intuitive experience, they do not achieve the broad and profound degree as the concept of *Dao*. The research into the *Dao* of technology is likely to become an important source for the development of modern philosophy of technology. Certainly, researches into the *Dao* of technology in the context of modern times need to establish an interaction relation between intuitive experience and logical analysis and can finally solve the following three questions:

(1) Conduct a holistic and dynamic analysis on the orgiastic relations between the internal and external elements of technological activities.
(2) Cultivate the sensibility on the trend of deviation against *Dao* of technology through intuitive experience, and reveal the reasons of deviation and the methods of reservation though logical analysis.
(3) Translate the conceptual understanding of *Dao* of technology into technological assessment and technological decision-making.

The further study will concern the epistemology, methodology, axiology and technological management of *Dao* of technology, which should be enriched and improved in the future.

References

Bunge, M. (1967). Toward a philosophy of technology. In C. Mitcham & R. Machey (Eds.), *Philosophy and technology* (pp. 62–76). The free press (1983).

Commoner, B. (1971). *The closing circle: Nature, man and technology* (p. 130). New York: Bantam Books, Inc.

Forgács, E. (1995). *The bauhaus idea and bauhaus politics*. Translated by J. Batki. Budapest, London New York: Central European University Press. See also Renke He. (1991). *The history of industrial design*. Beijing: Beijing Institute of Technology Press.

Haraway, D. (1991). A cyborg manifesto: Science, technology, and socialist-feminism in the late twentieth century. *Simians* (pp. 149–181). The Reinvention of Nature. New York; Routledge: Cyborgs and Women.

Heidegger, M. (1977). *The question concerning technology and other essays* (pp. 3–35). Translated by W. Lovitt. New York and London: Garland Publishing, INC.

Hughes, E. A. (1942). *Chinese philosophy in classical times* (pp. 185–186). London: J. M. Dent & Sons LTD. See also Confucius. *confucius analects: with selections from traditional commentaries*. Translated by Edward Slingerland. Indianapolis/Cambridge: Hackett Publishing Company, Inc. 2003. Tzu, C. (1994). *Wandering on the way: early taoist tales and parables of Chuang Tzu*. Translated by Victor H. Mair. A Bantam Book.

Jia, S. (1958). *Qin Min Yao Shu* (p. 30). Translated by Shi Shenghan. Beijing: Science Press.

Lao, T. (1999). Translated by Waley. Changsha and Beijing: Hunan People's (p. 99). Publishing House and Foreign Languages Press.

Liu, C. (1990). *Chinese system thinking* (pp. 551–552). Beijing: China Social Science Press.

Marcuse, H. (1964). *One-dimensional man: studies in the ideology of advanced industrial society*. Boston: Bacon Press. XI.

Mitcham, C. (1994). *Thinking through technology: the path between engineering and philosophy* (pp. 161–247). Chicago: The University of Chicago Press. See also Mitcham, C. & Schatzberg, E. (2009). Defining technology and the engineering sciences. In *Philosophy of technology and engineering sciences* (Vol 9). Edited by Anthonie Meijers.

Needham, J. (1954). *Science and civilization in China. Introductory orientations* (Vol. 1, pp. 3–4) Cambridge: Cambridge University Press.

Needham, J. (1956). *Science and Civilization in China* (Vol. 2, p. 228). History of Scientific Thought: Cambridge University Press.

Scott, D., Susan, M., & James, L. (2001). Design principles for ecological engineering. *Ecological Engineering, 18*(2), 201–210.

Shen, K. (1998). *Dream pool essays* (p. 555). Translated by Hu Daojing, Guiyang: Guizhou People's Publishing House.

Simon, R. (1951). Pursuit of happiness and lust for power in technological society. In C. Mitcham & R. Machey (Eds.), *Philosophy and Technology* (pp. 171–186) The Free Press (1983). See also Elull, J. (1963). The Technological Order. Ibid.

Van der Ryn, S., & Cowan, S. (1996). *Ecological design* (p. 18). Washington, D.C: Island Press.

Verbeek, P. P. (2006). Materializing morality: design ethics and technological mediation. *Science, Technology and Human Values, 31*(3), 361–380.

Verbeek, P. P. (2011). *Moralizing technology: understanding and designing the morality of things* (p. 91). Chicago and London: The University of Chicago Press.

Verbeek, P. P. & Slob, A. (2006). Technology and user behavior: an introduction. In *User behavior and technology development: shaping sustainable relations between consumers and technologies* (pp. 3–12). Berlin: Springer.

Wang, Q. (2005). From technique to Dao: The traditional philosophical idea of technology in China. *Philosophical Researches, 11,* 84–89.

Xie, S. (1989). *Heaven, man and image: Introduction of doctrine of Yin-Yang and five elements* (pp. 29–33). Jinan: Shandong Publishing House of Literature and Art.

Chapter 10
The Developing Process of Technological Rationality and Its Humanistic Relation

Jianjun Zhao

Abstract Technological rationality is a core concept in the philosophy of technology. Centered on this concept, varied researches conducted by scholars in different fields take on a multi-dimension and multi-level pattern, from employing the view of the Western humanism, technological pessimism or postmodernism, or basing on empirical or epistemological basis perspective, or exploring it by virtue of the Marxist thought on technology. This article divides the development of technological rationality into three stages, i.e. instrumental rationality, operation rationality and value rationality. It discusses the representative viewpoint in technological rationality critique, i.e. technological pessimism and hereby indicates technological rationality should be examined from humanistic perspective, and future technological rationality will be directed at the unity of implement rationality and value rationality.

Keywords Technological rationality · Technological pessimism · Humanism

10.1 Development Stages of Technological Rationality

In studying technological rationality, foreign researchers made an early start, marked by the research of Max Weber where rationality falls into instrumental rationality and value rationality. Herbert Marcuse, the representative of Frankfurt School first established the concept of technological rationality, and equated it with Max Weber's implement rationality (Marcuse, 1982). Subsequently, the Western humanists including Jürgen Habermas, Martin Heidegger and Lewis Mumford, postmodernists including Daniel Bell, David Ray Griffin, Alfred North Whitehead, Michel Foucault and Frederick Ferré, and modern technological critics including

J. Zhao (✉)
Department of Philosophy Teaching and Research, School of Central Committee of the Communist Party of China, No.100, Dayouzhuang, Haidian District, Beijing, China
e-mail: adad390@163.com

Q. Wang (ed.), *Chinese Philosophy of Technology*, Philosophy of Engineering and Technology 34, https://doi.org/10.1007/978-981-15-1952-9_10

Friedrich Rapp, Andrew Feenberg and Neo-Luddites also follow this tradition to further analyze and criticize technological rationality (Zhao, 2001).

However, it is thought that the widely accepted theoretical premise of H. Marcuse's definition of technological rationality needs to be questioned and reexamined. In Marcuse's opinion, technological rationality is instrumental rationality, which is the opposite of value rationality, and the strengthening of technological rationality must result in the weakening of value rationality. Based on the existing logic, it would not be strange to view technological rationality as the culprit of all kinds of problems of the modern industrial society and should be subjected to criticism, restraint and even denial. This logic has not only caused a shallow and one-sided theoretical perception of technological rationality, but also practical confusion.

10.1.1 Three Development Stages of Technological Rationalization

Following the approach of Marxist historical materialism, we discuss the rationalization process and its rich connotation of technology in modern society from a dynamic perspective. Based on its nature and forms, the development of technological rationality can be divided into three stages, i.e. instrumental rationality, operation rationality and value rationality.

(1) **Instrumental rationality**

Instrumental rationality, a kind of objective rationality opposite to value rationality, was first put forward by a German sociologist, Max Weber who investigated the Western modernity in the early twentieth century. In study of Weber on modernization, he describes the historical process of modernization as rationalization, and elaborates upon its structure and how it is manifested in the Western modern cultural concepts and institutions. Weber says "this kind of rationality is means-end rationality". The instrumentalization of rationality is due to the change of view of nature, namely the change from Aristotle's purposeful view of nature to modern science's mechanistic view of nature. This change loses the value of its existential objectivity and leaves it only subject to people's subjectivity. The world is a collection of matters and phenomena connected by causes and effects, and this change leads to the logical isolation of facts and values (Weber, 1994). Since value and meaning do not only exist in the objective world, but only is decided or created by personal or collective subjectivities. It has almost no objectivity and universality. The most typical functions of rationality are to reason and compute. Thus, the function of rationality can only be instrumental. The Western economics represents this typical instrumental rationality by making profits maximization as principles.

(2) **Operation rationality**

Operation rationality is a kind of procedural rationality that embodies and actualizes instrumental rationality. The meaning of operation rationality is to effectively achieve an anticipated goal by setting up a program or a plan through logic thinking regardless of whether it is ethical. For example, how to successfully clone a man is the only consideration in human cloning, no matter it is ethical or not. Operation rationality can provide the way of achieving the goal, but is not able to judge the goal.

The operationalism created by an American physicist Percy W. Bridgman in the 1920s reveals some characters of operation rationality. Operationalists regard the proposition and establishment of scientific concepts and theories as the results of operation, and therefore concepts that cannot be operationally analyzed are meaningless. The concept "operation" proposed by Bridgman mainly includes experimental operation, instrument operation and tool operation, which all come down to physical operation. Although mental operation and speech operation are also included, but mental operation submits only to physical operation. In this context, operationalism is the extension of pragmatism and positivism (Luo, 1995).

The operation rationality proposed here means a collection of operational programs, approaches, and modes that help to resolve the perplexing problems in order to achieve a certain goal and fulfill the rationality and purposiveness of implement by the guidance of instrumental rationality.

(3) **Value rationality**

Value rationality is a kind of subjective rationality opposite to instrumental rationality, manifested by the ultimate care and value guidance in social operant behaviors. Human's rationality tends to incessantly seek for the ultimate principle of value. The pursuit of this universal value entails the rationality and that is what is dubbed as value rationality. It regards value care as the core of human's value judges, adding subjective value selection into purely objective operation rationality.

10.1.2 The Contradiction between Instrumental Rationality and Value Rationality and Its Causes

When Max Weber divides rationality into instrumental rationality and value rationality, their cognitive opposition of subject and object manifests absolute isolation at the realistic level, namely the rationality of implement and the irrationality of value. instrumental rationality tries to reduce the universal validity of principle to the objectivity of law. To test this objectivity, the validity of practical application in natural science and engineering technology is employed as the criterion. It is evident that the great success of industrialization and technological revolution can be viewed as a clear proof of instrumental rationality's power. However, the striving to reduce the universal validity of value principle to the objectivity of a certain object has been proved to be logically impossible. It is because the appropriation in form is not equal to the value criterion, nor is the appropriation in value to effectiveness. This contradiction places modern society in dilemma in many situations, such as the contradiction between market economy and planned economy. Market economy is dynamic, but tends to dismiss moral value, causing the excessive pursuit of utility maximization. While, planned economy that to some extent can guarantee the appropriation of value could give rise to economic stagnation. Besides, the contradictions can also be seen between developing the economy and protecting the environment, between increasing efficiency and ensuring fairness, between improving welfare and spawning laziness, between strengthening medical care and social aging. The main reasons that account for "the rationality of implement and the irrationality of value" are as follows:

(1) **The difference of criteria**

Instrumental rationality follows the principle of the objective rule, pursuing regularity and economic utility, while value rationality is based on the sensible principle of subjectivity, pursuing harmony and existential satisfaction of subject.

(2) **A shift in empiricism**

Before the European Enlightenment Movement, empiricism underlying instrumental rationality's philosophy was once regarded as the opposite to rationalism. But here, empiricism is the only representative of rationality in the name of instrumental rationality.

(3) **The instrumentalization of operation rationality**

Operation rationality is a practice rationality that builds a connection between the subjectivity of value rationality and the objectivity of instrumental rationality. However, in realizing instrumental rationality, operation rationality only follows the instrumental comprehension of rationality, but fails to respond to the call of value.

(4) **The misplacement of technological means and human goals**

In modern technological society, human needs and goals have been identified with the technological goal of pursuing economic growth. In this case, with their practices in reaching their goals that turn into the purpose of technological activity, humans turn out to be conditioned to become means of achieving this purpose. Hence a complete break of instrumental rationality comes from value rationality.

Since Friedrich W. Nietzsche announced "God is dead", value nihilism has been increasingly prominent. As a result, the instrumental interpretation of rationality tends to be a basic thinking paradigm. The instrumental interpretation of rationality means conceptually denying the role of rationality in judging value. The great changes of modernization may witness instrumental rationality alone, making "age of rationality" almost become a synonymous of "age of science and technology". After the World War II, the technological determinism influenced all aspects of humanistic and social science. Technology is either publicized as the emancipated and constructive progressive force or denounced as the destructive force which leads to cultural crisis. Economic determinism always appears in the name of technological determinism, which determines what kind of value should be inherited through its relationship with technology. Values are not the inherent value criterion for judging human life, but viewed as the lubricant that serves instrumental rationality. That means rationality cannot function direct at values without the operation of instrumental rationality.

In this way, technological rationality regulates modern civilization, and technocracy emerges. Technological rationality increasingly shows senses of power politics and culture erosion, which means human spirit and society are entirely subject to technological thinking. In addition, the socialization of technology and the technicalization of society also help technological rationality dominate social development and human historical destiny. Meanwhile, misfortunes and disasters come along with human civilization. Heidegger said, "But where danger is, grows the saving power also." (Heidegger, 1977) In the 20th century, making reflections and criticism on technological rationality became a sacred mission of ideologists, among whom ideologists of humanism, Frankfurt School, ecology protectionism, the Club of Rome and postmodernism were the most prominent ones. In this process, there was a mental disposition that shows an increasing dissatisfaction with social reality, apprehension over modern science and technology with concerns about the future of human beings. This gradually forms a powerful philosophical trend, i.e. technological pessimism.

10.2 The Modern Manifestation of Technological Pessimism and Its Enlightenment

Technological pessimism is a term often used in recent years. Suspicion, denial and criticism on technology have a long history, such as Daoism's disdain for technique in ancient China and Rousseau's opinion "returning to nature" in the modern West. It is in the 20th century, these thoughts came to exert great influence on people's tendency of value behaviors as well as social choices. Technological pessimism employs a negative and irrational way of thinking to criticize technology, revealing the negative effects of technology controlling, repressing and enslaving human. The author demonstrates various manifestations of modern technological pessimism from the perspectives of rationality, humanism and real life.

10.2.1 The Modern Manifestation of Technological Pessimism

(1) **Criticism on rational level**

Hegel once advocated that rationality is the soul of the world, as well as the inherent, immanent and profound nature which constructs the world. Rationality is the interior quality that helps civilization rise and extend. Rationality is entitled to authority through its admiration for science and technology, and dominates the world in the name of technological rationality at the turn of 19 and 20th century. Criticism on the rationality level mainly reflected the criticism on technological rationality, and Max Weber, Herman Marcuse and Jacques Ellul are viewed to be the main representatives.

In the early 20th century, Weber divided rationality into instrumental rationality and value rationality. According to the differences in development of civilization, Weber believed that instrumental rationality with its increasing importance in capitalist society and its industrialization, would become the opposite of value rationality, and eventually phase it out to be the dominant rationality. Marcuse, the representative of Frankfurt School, regarded Weber's instrumental rationality as technological rationality, a derivative of instrumental rationality. According to Marcuse, technological rationality turns the rationality of techniques into the rationality of politics by the satisfaction of material wealth, and consequently becomes an effective way of controlling the society.

By luring people into misbelieving in some needs of no real necessity, it enslaves them, depriving them of their freedom. Technological advance means the expansion of slavery. As a result, man will be turned into one dimensional man with no critical mind or resisting spirit. Also, the society will be made one dimensional with no existence of its opponents or opposing voice.

If Weber and Marcuse criticize technological rationality from its external characteristics, then French scholar Ellul puts forward his doubts from the internal characteristic. In Ellul's opinion, modern society is technologically autonomous; and technological rationalization is a process which technological autonomy develops. It becomes a reality, self-sufficient with its own determinations on its accord. "Because of the autonomy of technique, modern man cannot choose his means any more than his ends." And technological autonomy is opposite to human nature. "The more technical activities are in society, the less autonomy and initiative people have", which means the effectiveness of technological rationality leads to the loss of freedom and value, and therefore the more powerful technological rationality is, the more disastrous loss people will suffer (Ellul, 1964).

(2) **Shout loudly from humanism**

During the first half of the 20th century, there are many scholars showing pessimism on civilization, technology and modernization within the perspective of humanism. Jacques Maritain was the first to express his concerns about human destiny in atomic age. In his view, the imprescriptible suffering has come when people grasped atomic technology. "The coming of the atomic age has suddenly exposed to the world the terrible countenance of this problem. Man no longer believes that science and technical skills can by themselves ensure the progress and happiness of the race. Rather, he is filled with terror at the sight of the destruction and calamities science and technical skills can bring about." (Maritain, 1952). The only path for human and the world to be survived are to waive science and technology, and go back to the religious world.

Oswald A. G. Spengler regards civilization as the inevitable destination of culture as well as the product of cultural collapse. Every advanced culture is a tragedy, and the root of this tragedy comes from the increasingly powerful and sophisticated scientific and technological activities. The technological splendor of the Western countries has not made people free from their sufferings, but worsened wars, exploitations, starvation and unemployment. These inexorably generate the disappearance of the Western world. He foretells that the western civilization will vanish in 2200.

Heidegger derived his questioning about technology from his long quest for its essence. "No matter how strongly we affirm or deny technology, we are always uncontrollably restricted to technology. However, the worst thing is that we have become technology's slaves when most of us readily accept that technology is neutral. And that is why we are blinded to its essence." (Luo, 1995). Heidegger interpreted technology as "framework", a way of manifestation. In the age of technology, humans are completely enslaved to the technological framework, and act upon the need of technology consciously or unconsciously. It is dangerous for human to take calculated thoughts as the only thinking activity and the utility of science and technology as the only standard to measure everything, for this will turn humans from "rational animal" into "technological animal". Heidegger emphasizes that technology is essentially beyond human's control. If human do not want to occur of the disasters brought by technology's violent movements, they should turn to themselves and get rid of the framework, which is almost impossible.

(3) **Conflicts at real survival**

In the 1960 and 1970s, promoting environmental movements were going on throughout the world as global issues emerged. Many scholars including scientists, philosophers, ecologists and writers condemned industrialization, modernization and science and technology. Typically, after the Club of Rome published the report *The Limits to Growth*, technology pessimism stepped out of academic, known by the public for its conflicts against modernization and industrialization.

The Limits to Growth is the first and the most influential report submitted by the Club of Rome. Through analyzing the most crucial parameters of human destiny, Meadows and other scholars drew the following conclusion, *that is*, the basic behavioral pattern of this world system is the exponential growth of population and resources, followed by its collapse. In order to avoid disasters, this report proposed the solution of "zero-growth economy" or "global equilibrium". The limits-to-growth proposition caused a huge storm all over the world. Although subsequent reports have continually revised its views, technological pessimism has already served as a banner. It has directly triggered the public condemnation and actions against the negative effects of technological application from ecological criticism, Greenpeace and even Green Party.

More notably, postmodernism, born in the latter half of the 20th century, shows a total tendency towards technological pessimism. As one of the modes of thinking that against modernity, it rejects modern technology and modern mechanistic world view based on it, criticizes subject-object dichotomy and anthropocentrism based on it, discards rationality and the resulting basis of realism, and uses "destroy", "dispel" and "game" to illustrate the meaning of life. David Griffin pointed out "we can and should leave modernity behind in fact, that we must if we are to avoid destroying ourselves and most of the life on our planet." (Griffin, 1988) Morris Berman believed "The subject-object dichotomy not only does not prevent but facilitate the unprecedented division, nihilism and the birth of destruction. Howard. P. Segal, editor of *Technology Pessimism and Postmodernism* proposes that no less important, technological pessimism has become an integral part of the emerging culture of postmodernism." (Ezrahi et al., 1994) Two international seminars organized by the US and Israel with topics on "technology and pessimism" and "technological pessimism and postmodernism" in 1979 and 1992, together with the appearance of Neo-Luddism, all of which reflected the modern rise of technological pessimism and people's great concern about it.

10.2.2 Several Enlightenments from Technological Pessimism

(1) Technological pessimism reflects the strong position of technology in the relationship between human and nature

Two significant factors are thought to affect the change of human-nature relationship. One is the range of human practices, and the other is the fulfillment of technological functions. The space and range of human practices mainly depend upon utilization of natural resources, more of adapting to nature, while the fulfillment of technological functions mainly depends on the improvement of people's cognitive abilities and the transition of the technology's role in intermediating in human-nature relationship, more of remaking nature. These two factors have already combined into one in modern civilized society. In other words, technology has become the determinant in modern human-nature relationship. With technology rising as a global power, it leads to the rise of artificial nature and technological sphere, and the overall expansion of technological society. The rise of artificial nature and technological sphere embodies the effect of technology and functions on human-nature relational structure, and the overall expansion of technological society embodies the effect of technology on human-society relational structure. On the one hand, all existing things are branded by the essential domination of modern technology. On the other hand, we cannot avoid that the two worlds we live in—the inherited biosphere and the created technological sphere—have lost their balance with conflicts. Therefore, the overall expansion of modern technology and the establishment of its dominant position result in the crisis and worsening of human-nature relationship. Under such background, technological pessimism appears and views seeking harmony between human and nature as its starting point and final goal.

(2) **Technological pessimism establishes the foundation of sustainable development theory**

It is known that global issues including global warming, population boom, species extinction, resource starvation and the development gap between the North and the South were initially proposed and studied by the Club of Rome. It is also the Club of Rome who was the first to disclose and criticize the disadvantages and potential danger of traditional development view which regards economic growth as the core. Although the solution "zero-growth economy" proposed by the Club of Rome is not feasible, the limited resources proposition, environmental value theory and social coordination theory are the theoretical basis of sustainable development theory.

(3) **Technological pessimism embodies the cultural traits of high-tech society**

F. Rapp reckoned technology as a complex phenomenon with both an application of natural force and a sociocultural process involved. This phenomenon can be studied from either technological or sociocultural perspective. Technology is a cultural phenomenon originally, and technological pessimism is a negative reflection on technological culture. Technological culture is human's examination and reflection on human activities. It uses technology as a "language" to reinterpret human's living environment and explain the evolutionary process of human. Thus, technological culture shows notable regional, national, contemporaneous, diverse and comprehensive characters.

Due to the relevance between technology and culture, technological crisis will inevitably cause cultural crisis, including belief crisis, moral decline and loss of personality and freedom. Technological pessimism reveals the distortion of technology on culture by criticizing it, making technological pessimism itself a philosophical cultural thought. The interpretation of technological pessimism should not be limited to its attitude towards technology, but regard it as a cultural phenomenon.

10.3 Humanistic Consideration of Technological Rationality and Its Future Fate

As discussed in the last part, people used to criticize technological rationality from the perspective of instrumental tendency of technology, and neglect the consideration on the cultural level. The analyses on the cultural level indicate that the conflict between human nature and physical property shows the contradictory relation between human and technology. Also, the humanistic significance of technological rationality is rooted in the internal connection of technological rationality and humanistic significance. And balancing the internal tension of technological rationality is the basis of coordinating human-nature relationship.

10.3.1 Human Nature and Physical Property: A Pair of Contradictions between Technology and Human Survival

In the real life, it is not difficult to notice that there are often various conflicts between technology and human survival as well as need and development, which can be regarded as the conflict between materialness and human nature after all. It often takes on the discord between unnaturalness of technology and naturalness of humanity. Undoubtedly, technology can cast great material effect on human. Technological means (such as machinery system) and products of technology (including its by-products, such as pollution discharge system) all belong to artificial objects—they come from nature but differ from natural objects, forming a dissident force in natural dynamic state system. When adding into the natural system as unnatural objects, artificial objects will impose unnatural disturbance on the natural material cycle, conversion and movements, destroy the original harmony and balance, and even cause a certain degree of disorder. The human body that has gone through a long-term natural evolution has the naturalness per se. When it receives a variety of unnatural technological products, such as artificial food, medicine and pollutant, it could bring about more or less the unnatural variation of human body, and even morbid damages when the intake exceeds the threshold value. But with the development of technology, people can manage to continually mitigate the detrimental effects made by technology on nature including that of human body. For example, gene therapy of some diseases shows this prospect compared with drug therapy. Moreover, people can also create expertise (such as environmentally-friendly technology) to make up for the destruction of nature caused by unavoidable industrial means. Humanity, making the human humanly, enjoys not only its naturalness, but also sociality, spirituality and other characteristics belonging to the higher level of humanity. This higher level of humanity is more likely to clash with the materialness of technology. For example, human get intertwined with tools when human want to use tool technology, and to some degree human (and humanity) should obey the tools (and materialness). This kind of obedience makes human suffer especially when humanity conflicts with physical property. By virtue of this, human feel the inhuman torture inflicted by tools with the emotional needs disregarded by physical movements, and human's freedom and dignity are wildly infringed to the point where human are even completely materialized.

Conflicts between human nature and physical property manifest conflicts between instrumental rationality and value rationality in the process of technological rationalization. "Capitalists use of machines" that Marx talks about has made workers trapped into a miserable situation. In order to reduce the negative effects of technology, on the one hand, technology should be constantly improved to be more consistent with humanity. For example, workers' physical sufferings are significantly relieved when automation techniques replace their direct operation on machines. Those sufferings will be aggravated if the attempt is given up or even is to turn backwards on technology "to the natural state". On the other hand, humans should be concerned and cared in the technological system, and their benefits should be protected from the system to the greatest extent. Moreover, aggravated damages of physical property of technology on humanity made by human should be minimized. It is not sufficient to identify one reason or make effort from a single perspective.

10.3.2 Technological Rationalization: The Process of Connecting with Humanity

From instrumental rationality regress to operation rationality, and then back to value rationality, the process of development process of technology rationality in some sense is the consideration of humanistic meaning of technological rationality and its humanistic relevance.

The rise of technological rationality started from advocacy for science and rationality in Renaissance. Renaissance has an inestimable influence on the emergence of modern science. As Eugeno Garin proposed, "If it is true that Humanism consisted in a renewed confidence in man and his possibilities and in an appreciation of man's activity in every possible sense, it is only fair to give Humanism credit for the new methods of scientific investigation, the renewed vision of the world and the new attitude towards objects with a view to using them and to dominating them." (Garin, 1965) Humanism has created a favorable moral and cultural atmosphere for science. Thus, science cannot be separated from the advantageous humanistic background. Under such circumstances, scientific rationality which has promoted science should have humanistic spirit, and technological rationality accordingly enjoys humanistic connotations.

In addition, the relevance between technological rationality and humanism also is shown in the technological rationality's contribution to human liberation. In the name of reviving ancient Greek and Roman civilization, humanist during Renaissance advocated science and democracy and promoted rationality and humanity. They opposed religious superstitions and engaged in human liberation. Since they have advocated science with the aim of human liberation, *that is*, to serve for their humanism theory, therefore, scientific rationality has an inseparable relationship with humanism. Most importantly, as the current result of development of human rationality, technological rationality is destined to own have the nature of human. Ancient Greek rationality is theoretical rationality which is used to explore mysterious nature, improve people's cognition of nature, and help people understand the relationship between human and the world. Rationality was regarded as the essence of human at that time. Not until the Renaissance when humanist came to exclaim against divinity and in favor of humanity, had the rationality recovered its nature of human. In modern times, scientific rationality develops along with science. Rationality is used to grasp the laws of world, requiring science to serve human in conquering and transforming the world, as well as creating abundant material civilization. In this stage, the function of rationality is more practical rather than understanding the world alone. In the wake of this practicability, technological rationality emerges and develops, which is the manifestation of human's "breakthrough willpower" to conquer the world. In short, no matter in what form rationality is, it is a fundamental element of human nature. Rationality's relevance to humanism provides possibility for modern people to reflect on technological rationality from the perspective of humanism.

10.3.3 Maintaining the Internal Tension of Technological Rationality: The Basis of Coordinating Human-Nature Relationship

Our reflection over technological rationality from the perspective of humanism means neither any elimination of technology, getting people back to natural state, nor an utter swing towards technology, leaving it holding sway over human being. It aims at maintaining a proper internal tension between instrumental property and value (humanistic) property when rationality works, therefore making nature and society develop harmoniously and people develop in a comfortable way.

Firstly, reflecting on technological rationality from the perspective of humanism is to seek a humanistic value that allows a reasonable scope where technological rationality transforms nature. This will help rebuild the harmony of human and nature. The attention to humanism does not mean advocating anthropocentrism, but coordinating the development of technological rationality with the meaning of human's being. To use technological rationality, human constantly bring their subjective initiative into full play on nature to acquire meaningful things for human lives and survivals. During this process, only by maintaining the harmony of human and nature can human's welfare be improved in that it is the only way that human can develop freely between society and nature. Anthropocentrism neglects the significance of this free relationship between human and nature for respecting human value for it emphasizes that human should be self-centered for his survival and development. On the human-centered basis, technological rationality is implement rationality which serves human with nature as its object to display its power. In this way, nature becomes an infinite container to bear the force of technological rationality. Therefore, reflecting on technological rationality from the perspective of humanism primarily requires the unification of human's naturalness and sociality. Human should respect law of nature when remaking nature. They should not only pay attention to the overall interests and ultimate value of human, but also admit the interest of nature and its internal value, and seek a harmonious development.

Secondly, we should correct the misconception of almighty rationality proposed by technological rationality, and restore human's original appearance by establishing a friendly cooperation between human and nature. In an age of scientism when almightiness of rationality was advocated, people believed that all natural, cultural, social and historical phenomena could be boiled down to dimension of science by a certain internal logical link. Thus, technology was regarded as an omnipotent instrument that could solve any problem. This conception lured people into believing they become means subjected to technology. Technological rationality vest in human originally, however, its omnipotence makes it an objective instrument that has infinite power with autonomy and regularity. As an independent external power, technological rationality governs human, making human attached to technology and causing human's alienation. Humans are no longer the master of technology, but, rather, become its slaves instead. As Marcuse states, human become "one dimensional man". Reflecting on technological rationality from the perspective of humanism at this level aims to recover the right relationship between human and technology by putting technological rationality under the guidance of value property, hence recovering the original character of human and establishing well-round interpersonal relationships.

10.3.4 The Future of Technological Rationality

With the background of modernity, systems as the representative of instrumental rationality seem to be all about the rationality. However, if dating back to ancient Greek philosophy, we can find that rationality mainly manifests as the exploration of rational foundation of the current value, in the view of Socrates and Plato, both of whom put value rationality in the first place of all the parts of human rationality.

Kant's practical rationality provided the basic rationality of ultimate value. In light of Kant's conception, human regularize their behaviors by moral, and in this way, they release themselves from the heteronomy of causations. By placing themselves in the realm of ends of human community, they can acquire the true freedom.

Max Weber distinguished instrumental rationality and value rationality, but he did not make them contradictory like Marcus. Instead, he tried to bridge their isolation. His greatness is that he has put forward the rational value of instrumental rationality. In his view, "modernity" is "rationality", and the "process of modernization" is the "process of rationalization".

It is believed that instrumental rationality is the initial phase of value rationality, while value rationality is the accomplished instrumental rationality. There is neither pure instrumental rationality that is value-free nor pure value rationality that is instrument-free. As two poles of technological rationalization, implement should develop to value rationality, and value rationality should be realized based on instrumental rationality.

In the historical process of modern civilization, there are always two strong trends of thought that are opposite and complementary. One is scientism which worships science and the makes progress and development as realistic orientation. The other is humanism which pursues spirit and takes value and meaning as the eternal theme.

If scientism and instrumental rationality are the driving system of modern society like engine and wheel, then humanism and value rationality are the control system like steering wheel and brake. If there is only the driving system without a control system, the society will stay stagnate or lose control (Shulman, 1996).

Value identification is the premise of technological rationality's manifestation. On the one hand, the flourish of cultural industries, such as automobile culture, screen culture and internet culture, are the cultural products of technological rationalization. It is the technological rationalization that makes cultures shown its diversity with high-tech, which reflects the real progress of society. On the other hand, the improvement of technological rationality from implement rationality to value rationality is of logical necessity for technology and is expected by culture as well.

10.4 Conclusion

By analyzing the inherent evolutionary mechanism and the value orientation of technological rationalization, this article points out that the current task is not to criticize or deny technological rationality, but to make great efforts to promote it and accomplish its transformation from implement rationality to value rationality. Therefore, regarding technological rationality as a developing rationality which contains implement rationality and value rationality, and meanwhile goes beyond them, is the key to solving realistic problems as well as a window to a bright prospect of the development of technological rationality.

References

Ellul, J.(1964). *The technological society* (p.134). New York: Vintage books.

Ezrahi, Y., Mendelsohn, E., & Segal, H. (1994). *Technology pessimism and postmodernism* (p. 3). Dordercht (Netherlands) Kluwer: Academic Publishers.

Garin, E. (1965). *Italian humanism; philosophy and civic life in the renaissance* (p. 221). Translated by Munz, P. New York: Harper and Row.

Griffin, D. (1988). *The reenchantment of science: postmodern proposals* (p. ix.). New York: State University of New York Press.

Heidegger, M. (1977). *The question concerning technology, and other essays* (p. 28). Translated by Lovit, W. New York and London: Garland Publishing.

Luo, J. (1995). *Opertionalism*. Y. Guangyuan (Ed.), *Encyclopedia of dialectics of nature* (p. 21). Beijing: Encyclopedia of China Publishing House.

Marcuse, H. et al. (1982). *The industrial revolution and the new left* (p. 82). Translated by Ren, L. Beijing: The Commercial Press.

Maritain, J. (1952). *The range of reason* (p. 178). New York: Charles Scribner's Sons.

Shulman, E. (1996). *Technology and the future*. Translated by Xiaobing, L. et al. (p. 60). Beijing: The Eastern Publishing House.

Weber, M. (1994). The christian ethic. In J. Habermas (Eds.), *The theory of communicative action* (p. 229). Boston: Beacon Press.

Zhao, J. (2001). *Inquiries into technological pessimism* (pp. 86–105). Shenyang: Northeast University Press.

Chapter 11
Three Dimensions of Nano-ethics

Sanhu Li

Abstract Nano-ethics is dedicated to exploring into the ethical and social significance of nanotechnology and currently has gone from the predication and control discourse of futurism and consequentialism into the dimension of practical exploration oriented by the reflexive co-evolution of technology and society. However, nanotechnology is to establish microscopic nano-space on atomic level after all and therefore it tends to be detached from man's intuition. This requires an integration of nanotechnology into human experience through the phenomenological strategy of visualization so that the optional flexibility of technology hermeneutics is offered to stakeholders. Meanwhile, technical operation on atomic level indicates the gathering of both intentional and non-intentional subject clusters on nano-scale. So, one systematic and stable development of multi-subject clusters should be realized ontologically, which contains construction of nano-world of natural science and engineering and accepts the participation of nano-world design of "natural social science". Only in this way, can we make nanotechnology socially and ethically healthy by the co-evolution of technology and society.

Keywords Nano-ethics · Practical exploration · Man's experience · Ontological governance

The prospect of nanotechnology, as a new field after the great breakthrough of information technology and genetic technology, has gradually stood out with the competition for political and economic interests or power among all nations ever since the end of last century. Various positive economic, social, cultural and political meanings have been attached to nanotechnology through the "revolution of nanotechnology" in the fields of science and technology, economy and policy. Meanwhile, ethical concerns about nanotechnology have also been taken into consideration in order to reflect on or respond to the attached positive meanings at

S. Li (✉)
Research Center for Urban Affairs, Guangzhou Academy of Governance,
No.99, Xianlie Middle Road, Guangzhou, China
e-mail: sh.li@foxmail.com

Q. Wang (ed.), *Chinese Philosophy of Technology*, Philosophy of Engineering and Technology 34, https://doi.org/10.1007/978-981-15-1952-9_11

the ethical level. At present, the ethical concerns about nanotechnology have been widely addressed. As nanotechnology philosophically involves man's existence and development in the first place, the name of "nano-ethics" has been widely accepted nowadays to refer to research on ethical and social significance of nanotechnology. Researchers in both foreign and domestic academic circles have reached agreement on the following two aspects in spite of various controversies in nano-ethics. Firstly, nanotechnology, as a newly rising field, must be evaluated ethically in the realistic sense though it is still a potential power in its universal application; secondly, nanotechnology must be considered as a technology strongly tied to social context, so as to reveal the absolute necessity of the dependence on nano-ethical evaluation. Based on these agreements, this article is to examine nano-ethics study from the three dimensions of practical exploration, human experience and ontology governance to display its process of deepening development and help people better understand the special position of nano-ethics.

11.1 Practical Exploration Dimension of Nano-ethics

From the perspective of man's practice, nanotechnology sprang out of man's universal technical attitude towards the world which treats miniaturized manipulation as the economic imperative in solving various problems under the guidance of technology. Nanotechnology is a rational way of dealing with nano-world and its connection with man. Its great breakthrough will not only reveal the nano-world to man but also prove man's existence in the nano-world. However, evaluating nanotechnology from the perspective of ethics is different from such macro approaches to industrial technology, information technology and biology technology, because nano-phenomenon is a kind of special and potential micro fact or event. If in the age of agriculture and animal husbandry, material products were produced with natural sources on the scale of chi or foot measurement; in the age of industry, large scale production was carried out by using natural force on the scale of micrometer measurement; and in the age of information, human communication is realized by using information resources and instruments on the scale of micron; then in the future nano-age, nature will be transformed with the fine and tiny material power as the carrier. That is to say, different from the technologies in previous ages which were confined to huge carrier of material energy, nanotechnology demands miniaturized manipulation of the nano-world.

The concept of "nanotechnology", rooted in the idea of material manipulation on atomic level put forward by Richard Feynman, an American physicist, was officially proposed by Eric Drexler, an American futurist, in 1986 and started to attract attention of the public. This history of nanotechnology naturally inspires ethical imagination of posthumanism, transhumanism about man's future to a great extent. Transhumanism includes various philosophical ideas including parahumanism, extropianism and singularitarianism, and so on. Posthumanism, which is the goal of transhumanism, is to bring man into posthuman age by newly arising technology.

Therefore, posthumanism includes transhumanism. FM-2030, an American futurist, writes in his book *Are You a Transhuman*? The transhuman is the earliest expression of some new evolutionary species. It draws supports from emerging technologies, ways of life and views of the world, and finally transforms to the so-called "posthuman". In 1990, Max More, an American philosopher, developed transhumanism into future philosophy with the above hypothesis as the theoretical foundation, and initiated a campaign of international culture or philosophy which supported the idea that emerging technology should be used to strengthen man's power and avoid man's deficiencies (such as disability, diseases, pain, aging and accidental death). Nick Bostrom, a British scholar, believes that transhuman is such an existence that possesses "a posthuman capability… over exceeding the maximum attainable by any current human being without recourse to new technological means" (Bostrom, 2008). He, as a posthuman futurist, believes in the promising future of biology technology, information technology and nanotechnology so much that he and his two coworkers recently decided to pay Alcor Life Extension Foundation for preserving their bodies by freezing them after their death, looking forward to the emergence of technological singularity that can call them back to life. Ray Kurzweil, an American posthuman futurist, believes that this technological singularity is not remote and that just in the year of 2045, it "will allow us to transcend these limitations of our biological bodies and brains. We will gain power over our fates. Our mortality will be in our own hands. We will be able to live as long as we want" (Kurzweil, 2005).

The predictions of posthuman futurism about technological singularity are the full imagination about nanotechnology which based on Drexler's conceiving of nano-assemblers (Drexler, 1981). Firstly, they predicted that the development of nano-assemblers will produce abundant material products for everyone's needs and that man's material demands are no longer a problem. Secondly, posthumanists predicted that nano-assemblers could become instrument-making machines that are programmable and able to produce multi-functional robots on the nano-scale. Thirdly, they predicted that there could be nano-robots that are able to enter human body, cure various diseases and postpone aging, so as to achieve longevity without any disease or even rest. Fourthly, they hoped that another type of nano-robots could re-design human body according to a perfect posthuman conception, and that other types of nano-robots could on the atom scale fulfill such tasks as brain activity replication, brain-computer interface uploaded by thoughts, perfect thoughts, super-smart, super-small and super-fast computers, and patients freezing to call them back to life.

Based on these predictions, the ethical promise of post humanism is an evolution will that promotes freedom, because current economic competition will make technology (nano-genetics, nano-robots etc.) and capital investors support each other and social Darwinism characterized by "survival of the fittest" has become the common ethical principle of evolution shared by both living creatures and technology.

Of course, nanotechnology cannot achieve its posthuman effects alone. Therefore, in the report *Converging Technologies for Improving Human*

Performance by National Science Foundation of United States, NBIC (nano-biology-information-cognition) is deemed to be converged on the nano-scale, by means of which human performance will be greatly enhanced (Roco & Bainbridge, 2002). Outside of human body, nano-information technology makes "bits and digital environment", and cognition-sociology changes "brains and memes and the social environment". Inside human body, nano-biology technology creates "bacteria and genes and the bio-environment", and nano-cognition science improves "bulk atoms, designed artifacts, and the physical environments". If it is believed that the development of nanotechnology and relevant artificial intelligence will help man overcome his own biological limitations, posthumanism tend to help man eliminate the ethical principle of social Darwinism. That is to say, neuro-ethics controlled by thoughts will replace gene-ethics controlled by selfish genes, achieving the ethical leap in the essence of life from struggle for life to a society governed by good and living up to the universal ethical promise of altruism.

The above two ethical promises of posthumanism seem to be contradictory to each other, but actually the two share the same moral judgment that nanotechnology is good, because it allows man to improve his abilities. This moral judgment is rooted within a virtue ethical tradition with a long history and also apparently related to consequentialism. However, consequentialism will not necessarily lead nanotechnology to goodness, as nanotechnology, judged from its result, is likely to be both morally good and bad. Posthumanism predicts or sets the potentially positive effects of nanotechnology only in the sense of the largest profit of the most people. Nevertheless, in the broad sense of consequentialism, nanotechnology risks man's health, freedom and environment. As to the latter aspect, ever since U.S carried out *National Nanotechnology Initiative* in 2000, not a few experts of technology and scholars of ethics have reflected upon the positive effects of nanotechnology from two perspectives. On the one hand, the positive significance of nanotechnology will not always be predicted accurately; on the other hand, nanotechnology itself cannot be separated from its poisonous feature. Just like posthumanism, this concern about its uncertainty and poisonous feature also stems from Drexler's conceive of nano-machines. While putting forward the concept of nano-robot, Drexler has been worried about the possibility that nano-robots may get out of control and reproduce themselves crazily, so that the resources on the earth will be turned into an enormous mass of grey goo consisting totally of nano-robots. However, Drexler's worry hasn't attracted any special attention of posthumanists. Only Bill Joy, the founder of Sun Microsystems, considers this material as a disaster to man, holding that the inborn threat of the knowledge about genetic engineering, nanotechnology and robots (GNR) is that its destructive capabilities of reproduction may result in bringing human development to an abrupt stop (Joy, 2000). Based on this conception, a world-wide ideological trend against nanotechnology has emerged. The Action Group on Erosion, Technology and Concentration (ETC), a radical environmental organization and a previous pioneer in anti-transgenetic food in Canada, released a report in 2006 which, by imitating the technological logic of Drexler and Joy, pointed out the combination of biology and nanotechnology would produce a new species of green goo, leading to the full scale destruction of environment, human health and biology variety. Nanotechnology is described in this technique as a field of knowledge that may destroy the world. This organization has not only persuaded the radical environmental protection organizations such as Greenpeace, but also attracted Prince Charles to advocate a stop of all nano-research all over the world.

Both the nano-ethical promise of posthumanism and the anti-nano ethical trend have considered both consequentialism and futurism. Their ethical propositions (such as human enhancement, grey goo etc.) are relatively novel, but they belong to the assumption or speculation of scientific fantasy. Alfred Nordmann, a British philosopher and Arie Rip, a Dutch scholar, criticize the nano-ethics which is confined to speculative discussions for being based on conditional chain of "if-then" straying from natural law, and using "the imagined future" to "overwhelm the present". This nano-ethics trend leads to a huge gap between the "oversupply" of nano-ethics and the "shortage of demand" for design and application of nanotechnology (Nordmann & Rip, 2009). In view of this situation, there has been a shift from futurism to the realistic concern about the poisonous feature of nanoparticles and environmental safety in nano-ethics. Among these concerns, Rip uses his Constructive Technology Assessment (CTA) method to put forward a framework of "reflexive co-evolution of nanotechnology and society" (Rip, 2005). This framework includes three concepts. The first one is "responsible innovation or development". We should consider the social orientation, expectation, intention and acceptability from the very beginning of the early development and innovation of nanotechnology. Innovators should be responsible for social actors and accept the enquiry of social actors so as to develop nanotechnology into an expandable system open to society. The second is "governance of new and emerging science and technology". Nanotechnology shouldn't be controlled from top to bottom but should be arranged on multi-levels, such as absorbing actors with the help of policy network system into taking part in and creating corresponding ways of interactivity, setting up projects of public fund etc., controlling and avoiding public disappointment by quality assessment of the future of technology. The third is the ethical, legal and social aspects (ELSA). ELSA has become a regular project of human genome research fund. Nanotechnology innovation fund project should also include ELSA so as to support "responsible innovation or development". According to this framework, it is neither necessary nor possible for CAT to assess the social impacts of nanotechnology that haven't occurred. Instead, assessors should interfere with the nano-community to accelerate the systemization of ELSA. In this way, responsible innovation will be supported by internal real time assessment, bridging of experts and amateurs, public involvement, introduction of the third party (such as insurance company), strategy game, and innovation competition etc. so as to promote the evolution of nanotechnology and social reflections. Rip has led his team Nano-Ned of Netherlands Research and Development Institute into cooperation with other institutes and done a lot of theoretical explorations into CTA in terms of norm, such as technology dynamics, governance theory, reflexive theory of co-evolution) and many other intensive and extensive experiential studies, such as strategic choice of nano-electronics, governance and arrangement of bio-nanotechnology.

The aforementioned CTA method differs from traditional technological assessment (TA) in that it timely influences technical design and practice by evaluating the uncertain ethical and social problems caused by technology. Traditional TA influences governance practice by evaluating social impacts of technology and therefore is hindsight of absolutism. However, it has to deal with the David Collingridge's controlling technology dilemma (Collingridge, 1980). One problem is the asymmetry of information. As long as new technology is not widely developed and adopted, it's not easy to predict its impacts. The other is the uncontrollability of technical power. As soon as new technology is widely used, it's difficult to control and change its social impacts. CTA has tried to get out of this dilemma by going beyond speculation. Instead, it adopts a kind of moral relativism to carry out empirical evaluation in the early stage of technology. The point is whether speculation is still needed for nano-ethics. Maybe, Collingridge's dilemma has been exaggerated. The difference between preventing beforehand and controlling afterwards is the purposes they have served. In this sense, Rip's CTA obviously belongs to applied ethics. While there's no doubt that speculative nano-ethics doesn't belong to applied ethics, but it involves significant philosophical themes such as the relationship between man and nature. As far as this theme is concerned, Amin Grunwald, a German expert in technical ethics, goes beyond the speculative category and has presented the current development of nano-ethics as "explorative philosophy of nanotechnology" (Grunwald, 2010). That is to say, nano-ethics has been down to earth and entered the field of practical exploration with realistic ethical concerns.

11.2 Human Experience Dimension of Nano-ethics

In the dimension of practical exploration, nano-ethics has put forward reflexive co-evolution of technology and society, which forms a sharp contrast with previous mindset of implementing ethical evaluation interference when problems occur. But from the perspective of epistemology, this ethical demand is to grasp nano-world intuitively. The rising of nanotechnology means building artificial nano-space on the atomic level. Nanotechnology tends to depart from the intuitive category of human beings. Therefore, Nordmann called it "noumenal technology", believing nanotechnology goes beyond man's touch, intuition and control. That is to say, nano-world is "a thing in itself" that hasn't been understood, experienced and controlled by man and its ethical dilemma lies in its breaking"the link between representation and control", and then making "the incommensurability" between the ethically controlled noumenal technologies and phenomenal (scientific) representations (Nordmann, 2005, 4–8). In order to make nano-world the object of man's experience, one phenomenology-hermeneutical method should be introduced into reflection upon nanotechnology, which has become a significant epistemological task in furthering nano-ethics study.

The phenomenological examination over nano-world is confined to proximity describing the objective nano-world. The related ethical consideration requires the presence of subjective intuitive nano-space. So, what nano-scientists and engineers are doing is “to promote the strategic development of visualized images at nano-space” (Li, 2009, 87, 89). The scientists of IBM joined 35 xenon atoms into “IBM” (The picture first appeared in *344vol. 1990 Nature*) with scanning tunneling microscope (STM). The entire space is less broad than 3 nanometers. In spite of the fact that it’s not important for nano-scientists and nano-engineers to imagine the length of nanometer, the length of nano-world indeed has gone beyond man’s intuitive imagination. On International Congress on Imaging Science held in Amsterdam, Netherlands, Dec. 7, 2004, Don Eigler from IBM pointed out that “if you can imagine anything that’s a billionth of anything else, you are way of ahead of me”. The question deserving careful examination is to what extent the image of nanostructure obtained by scanning tunneling microscope has reflected the accurate presentation of what makes nano-world as it is. As far as its relationship with man’s experience, there exist three worlds in nano-world. The first one is the nano-world as thing-in-itself, including technically artificial products on the nano-scale, which makes up the transcendental control over man. The second one is the nano-world that can go into the horizon of man’s experience, including statistic graphs, technical images and so on. The third one is the nano-world that can be grasped rationally, such as the mathematical model of nanoparticles. Apparently, the first one is farther away from man’s intuitive and empirical world than the other two. Since man cannot directly approach the first level of the nano-world that he tries to present, he can only understand its phenomenal content through the second and third levels, especially technical images. As pointed out by Joseph Pitt, an American technology philosopher, “The problem is not that we do not know how to interpret what is presented to us as an image, but, rather, that we have loaded the creation of the representation ahead of time without being able to know if our guess that this is what the STM and its accompany computer programs are producing is an accurate picture of what is really there” (Pitt, 2005, 31). That is to say, nanotechnology as “noumenal technology” stands for the trend of invisible or miniaturization of the artificial world’s surpassing intuition. As far as the invisible nanotechnology is concerned, the epistemological challenge is how to promote the strategic development of visualizing the nano-world so that man can propose and discuss questions of the relationship between nanotechnology and ethics in the intuitive world.

2004 was the year of nano-phenomenology, when three conferences were held focusing on nano-technical images. Apart from the Amsterdam conference held in December, the conference “Imaging and Imagining Nano-science and Nano-engineering” was held in University of South Carolina in March, 2004, and Scientific Imaging Conference in Berlin, Germany in December. In May 2005, Center for Interdisciplinary Research in Bielefeld University sponsored a conference focusing on nano-space imaging. The theme of all these conferences was to visualize the nano-world by the codification of nano-system. Thomas Staley says there are three choices in the development of nanotechnology visualization (Staley, 2008, 16). The first is that a limitation lies in man's approaching the nano-world. The gap between the nano-world and the mesoscale world is so wide that it cannot be bridged, which makes virtual atomic experience impossible. The second is to combine the relatively few functional experiences of the nano-world with the phenomenological imagination to establish appropriate nano-space models and to describe the nano-world as objectively as possible. The third insists that it is possible to approach the nano-world phenomenally. Nevertheless, the understandings of the nano-world and the normal experiences are incommensurable or imbalanced because of the different information in these two worlds. The first choice leads to nothing, which leaves the other two for further research. For example, based on quantum corral, Eigler and Nicholas Brodbent have shown the constructive abilities of different technical imaging methods by orange corral and blue corral respectively. These two technical imaging methods are nearly the same but with the inclination to go beyond technology itself. Orange corral, aimed at obvious image concept, is combined with personal choices, communicating desire and subjective touch experience. It tries to describe one-way visual image like IBM. However, this technical image, just like Démon de Laplace or Maxwell's demon, is closer to a scientific hypothesis or technical explanation of determinism, than to phenomenological contact with the substantial nano-world. Comparatively speaking, blue corral is more in agreement with the image of virtual reality and its coded color program is helpful for the direction location of man's eyes so as to achieve the original image emphasized by colors. Although Eigler emphasizes the importance of communicating with the nano-world personally as an observer (namely the second choice), it's not a direct specific observation over atomic phenomenon but an empirical participation into the nano-world. Therefore, Staley believes we should focus on the virtual reality of the nano-world as Brodbent did (namely the third choice) rather than over-advocate communicating personally with the nano-world archetype.

Currently, in spite of the incommensurability or asymmetry among all imaging treatments of nanotechnology and the nano-world, these treatments can integrate nanotechnology into the category of man's experience, and present the flexibility of phenomenology-ethics-hermeneutics interpretations of nanotechnology. Once nanotechnology is approached from the perspective of hermeneutics, philosophers can construct the route of nanotechnology in a broader sense, which is exactly what Martin Meyer and other researchers have done. They have divided the future development of nanotechnology into 6 categories according to time (present and future) and technology (the specific technology and the aim): from realistic technologies to future technologies (scanning tunneling microscope), from realistic technologies to future aims (the development of abundant nano-materials) and from future aims to future technologies (designing the ultra-thin-layer involves many future aims that are encouraged by current technologies, which makes up the technical starting point of future designs of the ultra-thin layer), from future technologies to future aims (the nano-machine hypothesis by Drexler), from realistic aims to future technologies (Smalley's method of biology imitation) and from realistic aims to future aims (Meyer & Kuusi, 2004, 157). Realizing the last one need depend on explaining users' demands, but which is lack of technical foundations. The other five, as the specific illustrations of nanotechnology, basically reflect the process of the development of nanotechnology. The first three all have excellent technical foundations and belong to the category of realism. The second one in particular is the realistic foundation of criticizing and envisioning nanotechnology. The last two obviously belong to the category of non-realism. The difference between the two lies in that the fifth one can interpret modern technology and set future technical aims while the fourth one is basically a scientific hypothesis. Considering the Drexler-Smalley controversy, the fourth and fifth ones are actually two optional paradigms of nanotechnology.

In the sense of science, Drexler's hypothesis of nano-machines and Smalley's method of biology imitation are two fundamentally different and incommensurable ways of molecule assembly. The former one, although based on the engineering archetype, explores the possible ways of molecule assembly from the perspective of mechanism and pure theory. Its essence is to establish a theoretical model that doesn't violate physical principles. It adopts the bottom-up method. While Smalley's method of biology imitation, rooted in detectable or controllable effects, emphasizes the practical adaptation of chemical structures and uses the top-down method. The incommensurability between the two only indicates that the standards set for one method are not applicable to the other. However, that doesn't mean we cannot compare the theories, concepts, methods and tools used in the two methods. If approaching the practices of nanotechnology accordingly, this controversy can be dealt with by considering the instrumental relationships in the first and second categories. It is in the instrumental sense that the two can achieve the "minimal agreement" or "minimal fusion of horizon". To Drexler, excluding instrumental requirements from theoretical methods equals to ignoring the key problem of nano-machines' adaptation to nano-space. While to Smalley, insisting on the controllability and detectability is to emphasize proper instrumental requirements. Based on this instrumental requirement, Otávio Bueno points out on the Drexler-Smalley controversy, "after all, both parties share their commitment to the indispensability of appropriate instruments to control nano-phenomena". But Smalley has introduced an instrumental requirement in order to make sure that molecule assemblers are possible "at least in principle, to produce detectable results", so "the overall proposal Smalley advocates seems more adequate" (Bueno, 2004). Though this doesn't mean Smalley's plan is completely applicable, the emphasis on instrumental requirements requires explaining various controversies about nanotechnology by integrating them into the real world, so as to show the function of various methods in constructing nanotechnology. The importance of this instrumental relationship is shown in two aspects. In the first place, instruments should be adopted to understand atomic and molecular phenomena at the theoretical level and to construct images of the nano-world, promote the stabilization of nano-products and establish models of the nano-world at the experimental level so as to benefit scientific imagination. Secondly, if the nano-world shown by experimental instruments is not a result of the absolute yardstick, the nano-world shown by the instrumental relationship can only become the significant source of explaining the noumenal participation of nanotechnology. That is to say, we cannot understand and control the nano-world from the perspectives of positivism, objectivism, reductionism, mechanism and determinism. Nor can we consider nanotechnology as simply an objective description or manipulation of the microscopic physical space. Instead, it is a hermeneutical activity open to aesthetics, sociology, ethics and even politics. Only in this way, can we grasp the structure-function of the nano-world as a whole and its wide social and ethical significance so as to avoid deciding man's future with single determinism of miniaturized mechanism (Li, 2009, 91).

11.3 Ontological Governance Dimension of Nano-ethics

The practical exploration of nano-ethics at the epistemology level is not confined to phenomenology-hermeneutics activities. Its development inevitably gives prominence to nano-ethics in scientific interpretation. In this sense, not a few scholars recently refer to the nanotechnology related academic fields as "nano-technoscience". Nano-scientists consider nanotechnology as understanding and controlling materials on the nano-scale which is above 10^{-9} meter. On this scale, some materials may present some particular qualities so as to produce corresponding products. Both the practical explorations of nano-ethics and the dimension of man's experience predict and reflect upon technical manipulations on this scale. In this sense, nano-scale naturally becomes a core issue in understanding nanotechnology.

If nanotechnology is not simplified as an unknown field, it will be an important epistemological task in nano-ethics to understand nano-scale from the perspective of scientific theory. Nordmann, in the name of "nano-technoscience", puts forward "closed theory" to understand nano-scale problems. The so called "closed theories" refer to Newton's theory of classical mechanics, Maxwell's electromagnetic theory, thermodynamics, statistical mechanics, non-relativisitic quantum mechanics and so on. These theories have all achieved their ultimate historic forms, applicable scopes, critical immunity and permanent appropriateness. Nordmann therefore believes studies of nano-scale have three characteristics (Nordmann, 2008, 221). Firstly, there are fundamental differences between classical mechanics and quantum mechanics. Classic theory is unable to describe quantum phenomena, and quantum theory is not suitable to describe classic phenomena. Secondly, nano-scale phenomena include purports in knowledge and technology, which is an exotic field to both the classic properties of matter above the nano-scale (such as color, electrical conductivity etc.) and the quantized phenomena under the nano-scale (such as quantized electric conductance). Thirdly, researchers of the nano-scale can only appeal to large-scale theoretical packages from quantum and classic mechanics to scientifically explain the innovative characteristics, behaviors or processes of nano-products. This means nanotechnology cannot be described, and scientifically understood by using any single closed theories. Instead, both closed theories need to be adopted at the same time to interpret nanotechnology. Now people describe the nano-world as a complicated, self-organizing, bizarre chemical-biological world, emphasizing on the production of active nano-machines. However, the current achievements of nanotechnology are confined to the production of new nano-materials (passive nanostructures). Active nanostructures or systems can only be produced by integrating molecular activities into the nano-technical system. Considering closed theories related to nano-scale, people haven't witnessed the birth of any magical active nanostructures, except for corresponding theoretical planning and formalization. Although the descriptive terms for the nano-scale phenomena like "selective surface", "the surface self-cleaning", "smart material", "self-movement" and "self-expansion material", carry some special meanings, theoretically they are merely metaphors drawn from the outside of the nano-world instead of realistic active nano-systems. That is to say, we cannot observe the special complexity of nano-scale phenomena from closed theories related to nano-scale. Instead, we can find that we don't express difficulties in understanding and controlling nano-scale phenomenon (Nordmann, 2008, 223). This shows there exist some blind areas in understanding nanotechnology and that more disciplines should be involved to further our understanding and controlling of nanotechnology.

Though explanations offered by closed theories related to nano-scale cannot overcome the difficulty in understanding and controlling nano-scale phenomena, the closed characteristics of those theories need the presence of multi-theoretical explanations of nanotechnology. This means we cannot deal with nano-scale problems with a philosophical notion of reductionism; instead, we can consider it as an interdisciplinary area. Bernadette Bensaude-Vincent wrote in the editor's note for the column of philosophy of nanotechnology, *HYLE* (First issue, 2009), a magazine of international philosophy of chemistry, that nanotechnology had blurred the boundaries of many disciplines. This blurring is presented as the difficulty to distinguish between nano-science and nanotechnology, whether the research on nanotechnology is aimed at something practical or epistemological, such fundamental material units as atoms, molecules and macromolecules are all considered as the functional units of devices, engines or mechanics. This blurring is also shown in the converging technology of NBIC, which attempt to eliminate the boundaries between man and machines, nature and artificial things. It seems that "bio is nano". Bioengineers use cloning and synthetic biology to "cross the boundary between the products of life and the products of human technologies", and even make living matter "instrumentalized for technological purposes" (Bensaude-Vincent, 2009). Just because of this, disciplines like nano-biology technology, nano-chemistry technology have emerged.

Different from studies on the disciplinary boundaries at the micro level, Vitaly G. Gorokhov recently has defined nano-technoscience as "a combination of natural and engineering sciences" (Gorokhov, 2012) at the macro level. Nano-technoscience on the one hand is similar to classic natural science in its explanation of natural phenomena and prediction of natural events with mathematical and experimental data, and on the other hand is close to engineering science in its designing for new experimental situations and planning unknown new nano-systematic structures in nature or technology. The process from nano-theoretical computation to nanostructure planning involves parameter calculation, mathematical modeling and engineering function planning as well. That is to say, engineering problems should be transformed into scientific issues and then further transformed into mathematical issues that can be solved by deduction. Current research on nano-scale is done either by means of bottom-up (molecular assembly on micro level) or top-down (chemical synthesis or biological imitation on macro level). The former stands for planning and analyzing nano-theory and the latter stands for the synthesis of engineering planning. As the combination of natural science and engineering science, nano-technoscience is a systematic project. Gorokhov mentions the binary structure-functional feature of nano artificial products methodologically. Similarly, Mare J. de Vries adopts the method of structure-function to discuss the border issues in research on nano-scale in a broader sense (de Vries, 2005). Vries, analyzes abundant factors related to nano-research based on non-intentionality (structure), intentionality (function) and comprehensiveness (structure-function combination). These factors involve mathematics, physics, chemistry, biology, psychology, logics, culturology, sociology, economics, aesthetics, law, ethics and religious studies, which shows the multidisciplinary feature of nanotechnology. This allows philosophy of nano-science to be combined with nano-engineering, nano-science, nano phenomena manipulation and control to enable nano-community to better understand and control nano-technical practices.

Both nano-scale studies and disciplinary boundary studies have displayed the heterogenicity of nanotechnology. As aforementioned, various approaches to nanotechnology exist even if the concept of technological paradigm is adopted. Confronted with this heterogenicity, nano-ethics must appeal to the concept of nano-techscience to ontologically achieve the holistic understanding and even control of nanotechnology. Meyer and some other researchers consider nanotechnology as a sort of paradigm system so as to include various approaches to studying nano-scale. The differences in those approaches can be distinguished by means of epistemic utility (Meyer & Kuusi, 2004, 165). Epistemic utility refers to the predicative effects of various technological paradigms, the positive and negative values given to those effects by stakeholders, the technological means to adapt to various technological paradigms and the effectiveness of the related evaluations, mainly the evaluation of nano-technical means and its influence to highlight the intention priority of the paradigm system of nanotechnology. Meanwhile, not a few scholars have adopted such concepts as knowledge graph, subject cluster, topics network to endow nanotechnology with an academic position of "super ontology", because nanotechnology indeed has reached its peak in involving various intellectual fields of human beings. People emphasize the mutual relationships among the related disciplines in the sense of system theory, which partially covers the meaning of holism. However, if Vries' structure-function method is adopted, nano-techscience could be divided into two subject clusters, namely, the non-intentional topics network, with both fundamental research such as biology, chemistry, physics, mathematics, engineering, calculation and industrial application and specific technology such as nanostructure, nano-equipment and nano-electronics, and the intentional network, including toxicology, environment science, psychology, policy science, economics, sociology, culture science, ethics and so on. The problem of this dichotomy is that in the sense of nano-technoscience, people always intend to consider the former as ontological subject cluster, and the latter as subsidiary subject cluster, which gives rise to the priority of non-intention. In order to solve this problem, Michiko Tanaka, using the popular top-down method, offers a kind of meta-level ontology for nano-scale study which considers nano-science and nanotechnology as the top meta-level concepts, and nanophysics, nano-chemistry, nano-mateiral science, nano-biomedical science, and nano-technical engineering as the bottom meta-level concepts (Tanaka, 2005). This division is more a pursuit of the priority of non-intentional subject as ontology than an adaptation to ontological demand of governing by subject division or philology and hence put nano-ethics in the non-priority position.

The above ontological controversy over intention and non-intention is fundamentally about whether nanotechnology is basic or applied science. Actually, nano-researchers are both nano-scientists and nano-technicians. Dialectically speaking, nanotechnology is neither basic nor applied science. In the past, basic science and applied science were seen as the description, understanding and reformation of nature, or the artificialization of natural things. However, nanotechnology has already gone far beyond this definition to pursue the naturalization of artificial things (Drexler's design of nano-assemblers is aimed at this goal) and become so-called "real technology". Xavier Guchet therefore believes that what nanotechnology has conveyed is a kind of "operational view of nature". Therefore, nanotechnology should be considered as the "science of operations" (Guchet, 2009). To this science of operation, the distinction between natural and artificial things is pointless, because the ultimate form of nanotechnology can eliminate any original natural structure according to naturalizing artificial things. There is no more distinction between nature and man, man and machine. Further, the distinction between structure and function, means and end is no longer meaningful, as this distinction cannot resolve such ontological problems as what exists, what doesn't exist, and what is the origin, what is the derivation. Peter Galison calls this "ontological indifference" of nano-technoscience (Galison, 2006). For example, researchers of molectronics needn't make sure weather electrons exist or not while they pay more attention to imaging explanation of electron transmission and use such concepts as electronic orbit, electronic wave function and so on, to describe the property of non-material probability of the quantum world. It is by means of this privilege of ontological indifference that researchers of molectronics have developed quantum mechanical descriptions of the complexity and delicacy of electronic transmission. Similarly, researchers of nano-scale did not produce "domesticated objects" or "substitutes" of nature in laboratory by objective reduction of natural objects through relevant technical means. This ontological indifference of nano-science is understood as "disinterestedness" by researchers. Therefore, nano-scholars' descriptions of the nano-world are nothing more than instrumental rationality explanations and hermeneutical flexibility in some other senses in ethics. Therefore, the so-called ontological indifferences need to be understood with multiple hermeneutical meanings. Nordmann believes that understanding and controlling nanotechnology should be focused on the powerful potentiality of imaging the nano-world, transmitting from pursing "cognition certainty" to realizing "system stability" of multi-subject cluster. This includes both establishing the nano-world in natural science and engineering and designing the nano-world in the social science of nature so as to develop nano-technoscience into a "social sound technology" (Nordmann, 2008, 16–20). In this sense, it is by means of ontological governance and co-evolution of technology and society that nano-ethics tries to realize the goal of sound development of nanotechnology.

11.4 Conclusion

With the above discussion, let's reconsider the attitude towards nanotechnology. Currently, scientists, engineers and even decision makers have confidence in the various positive meanings of nanotechnology, because they hold such an attitude that nanotechnology can reveal the truth and reality of the nano-world. According to this technical attitude, man can not only understand but also reform the nano-world. So this attitude is called objectivism, positivism, reductionism, mechanism etc. (Li, 2001). However, there are two problems with this attitude. Firstly, the ontological method of bottom-up requires nano-science to control over the nano-world by manipulating the atomic and molecular structures. It is inevitably bound to the instability of the nano-world and technical equipment, which may lead to butterfly effects that may result in tremendous change out of tiny disturbance. Secondly, nano-engineering pays attention to the causal dependence of knowledge, action and control, technically exaggerating that the nano-world is omnipotent. It obviously ignores the participant influence of the mesoscale world over the nano-world as shown by modern nonlinear physics, which will lead to the "fat fingers problem" and "sticky fingers problem" (Schmidt, 2004) that nano-machines established by atoms cannot operate atoms. It is not realistic at all that nano-science is to establish an objective nano-world excluding man. In other words, the development, application and dissemination of nanotechnology cannot create the specific tiny physical time and space without man's participation.

Since man cannot approach the nano-world completely at the epistemological level, he can only promise to close or minimize the huge gap between the uncertain nano-world and the world that man lives in as much as possible. Nanotechnology, as the bridge between man and the nano-world, needs to consider "nearness" as the basic principle of nano-image communication. This nearness is not to sheer overcome the special physical, chemical and biological difficulties. It is also an ethical need to regard man as the other of the nano-world. Nano-technical imaging is a kind of presentation,a specific outcome of man's grasping prominent features of objects, scenery, situation of things and spiritual notions. What man uses to describe the prominent features of the nano-world becomes the function or intention of man's constructing nano-technical imaging. Therefore, Feynman's "plenty of room at the bottom", Drexler's automatized nano-machine and Eigler's atomic operation under scanning tunneling microscope are not instrumentally exploring and reforming the microspace in a way independent from man, but a further promotion of the relationship between man and the world. In view of this, rather than understand and control the nano-world in the sense of ontology, objectivism, reductionism and mechanism, we should take an functionist attitude of structure and property to reveal utility (Staley, 2008, 22). To put it another way, we integrate various angles in dealing with the nano-world by means of holistic metaphor to produce and negotiate its various possible meanings in the sense of public culture.

The above shift of technological attitude shows that, with the broad meanings of nanotechnology, we can reflect on nanotechnology socially and ethically from the relationships between man and technology, between man and man. As to the first relationship, if nanotechnology (science) is not to grasp the absolute substance of the nano-world, its mission is to bridge the invisible image of the nano-world and man. Man acquires a better understanding of the nano-world by coding the technological images of the nano-world. But this doesn't exclude the possibility that some scientists and engineers may go beyond the current "limitation of knowledge" in the rational sense to the "moral ambition" or "enthusiastic propaganda" that nanotechnology can omnipotently control the nano-world. Therefore, we need to consider nanotechnology as "noumenal technology" in the sense of morality, confining it to man's rationality, responsibility and control. This "freely standing" nanotechnology is sure based on man's awe and veneration towards the nano-world. It requires rational calculation, domestication, conquering and logic of control to be performed with man's intuition and sense of responsibility (Nordmann, 2005, 18). The second relationship is involved right here, namely the technological relationship between scientists or engineers and common people. Nano-scientists and nano-engineers publicize their scientific narration and technological designs to common people in public sphere. Although they mainly appeal to grasping the truth and describing the substance of the nano-world, their renderings win enthusiasm and even trust from common people because of not only the truth and entity itself but also their constant effective power and senses of responsibility.

Once this issue is considered in terms of public sphere, people will go beyond the current rational knowledge of nanotechnology to various social, political and ethical imaginations about nanotechnology's potential of changing man's ways of living. Under the circumstances that man has not known the real nano-world, the following determinism is prevailing nowadays-the assertion that current nanotechnology image is the standard model or the real presentation of the nano-world, and that tremendous success will be achieved to reform the nano-world in this way. This way is obviously misleading to common people. Confronting with this complicated nano-world, people may be led "into trouble at the political level and it should get us into equally big trouble in our epistemology" (Pitt, 2005, 32).

This dilemma or trouble includes two contradictory dimensions. The first one is the Prometheus-stealing-fire typed declaration that auspicious nanotechnology, representing a new myth of Houyi shooting the sun or Goddess Chang Er flying to the moon, will change man's fate with its tremendous power by continuously fulfilling man's needs, wishes and aspirations. The other one is the inauspicious prediction that nanotechnology stands for a fable warning the world. According to this fable, nanotechnology, as a new Démon de Laplace or Maxwell's demon, is nothing but the technological egomaniac building the Babel Tower in *Holy Bible* and the Frankenstein in Mary Shelley's novel who ruined himself because of his Jekyll-Hyde-ish dual personality. It would lead man to doom in the end as Kuafu did to arrogantly chase the sun. From the perspective of nano-ethics, the ultimate care of these public impressions or comments is to control actions with rationality. That is to say, the development, application and dissemination of nanotechnology must be rooted in the presupposition of the meanings of man's community that nanotechnology cannot simply become man's background or environment (technical environment). Rather, nanotechnology should be regarded as the focus or the perceptive extension of man to seek social perception and moral reciprocity from transcendental ethical nearness so as to perpetuate the various symbolic meanings between man and nano-world, such as tolerance, care, promise, place, body, situations and so on. It is the charm of the virtual nano-world that the signal communication between man and the nano-world not been a technological background or "Gestell" which imposes on man. Instead, nanotechnology is regarded as a controllable medium of communication between man and the nano-world through ontological governance, focusing on man's empirical intuition and perception and based on phenomenological-hermeneutical flexibility. Even though this prospect of nano-ethics is not a feasible reality, it should at least be taken consideration into the presupposition of nanotechnology revolution as the moral intention.

References

Bensaude-Vincent, B. (2009). Boundary issues in bionanotechnology: editorial introduction. *HYLE, 15*(1), 2.

Bostrom, N. (2008). Why I Want to be a posthuman When I Grow Up. In B. Gordijn & R. Chadwick (Eds.), *Medical Enhancement and posthumanity* (p. 108). Springer.

Bueno, O. (2004). The drexler-smalley debate on nanotechnology: incommensurability at work. *HYLE, 10*(2), 95.

Collingridge, D. (1980). *The social control of technology* (p. 11). New York: St. Martin's Press; London: Pinter.

de Vries, M. J. (2005). Analyzing the complexity of nanotechnology. *Techné* (Vol. 8, Issue 3). Spring.

Drexler, K. E. (1981). Molecular engineering: an approach to the development of general capabilities for molecular manipulation. *Proceedings of the National Academy of Sciences, 78* (9), 5275–5278.

Galison, P. (2006). The pyramid and the ring. In: *Presentation at the Conference of the Gesellschaft für analytische Philosophie (GAP).* Berlin.

Gorokhov, V. G. (2012, April 4). Nanotechnoscience as combination of the natural and engineering sciences. *Philosophy Study* (Vol. 2).

Grunwald, A. (2010). *From speculative nanoethics to explorative philosophy of nano-technology* (p. 4). No: Nanoethics.

Guchet, X. (2009). Nature and artifact in nanotechnologies. *HYLE, 15*, 1.

Joy, B. (2000). Why the future Doesn't Need Us. *Wired, 8*(4), 2000.

Kurzweil, R. (2005). *The singularity is near: when humans transcend biology* (p. 9). New York: Penguin.

Li, S. (2001) *Small world and big outcome: the nano philosophy towards future* (pp. 72–83). Beijing: China Social Science Press.

Li, S. (2009). Nano-phenomenology: Image explanation and intention ethics of construction in micro space. *Philosophical Researches*, *7*(87), 89.

Meyer, M., & Kuusi, O. (2004). Nanotechnology: Generalizations in an interdisciplinary field of science and technology. *HYLE, 10*(2), 157.

Nordmann, A. (2005). Noumenal technology: reflections on the incredible tininess of nano. *Techné*: *Research in Philosophy and Technology* (Vol. 8, No. 3, pp. 4–8). Spring.

Nordmann, A. (2008). *Philosophy of Nanotechnoscience*. In G. Schmid, H. Krug, R. Waser, V. Vogel, H. Fuchs, M. Grätzel, K. Kalyanasundaram, L. Chi (Eds.), *Nanotechnology (*Vol. 1, p. 221). Principles and Fundamentals, Weinheim: Wiley.

Nordmann, A., & Rip, A. (2009). Mind the gap revisited. *Nature Nanotechnology, 4,* 273.

Pitt, J. (2005). When is an Image not an image. *Techné, 8*(3), 31.

Rip, A. (2005, January 13–15). Technology assessment as part of the co-evolution of nanotechnology and society: The thrust of the TA program in NanoNed. In *Paper Contributed to the Conference on "Nanotechnology in Science, Economy and Society"*, Marburg.

Roco, M., & Bainbridge, W. S. (Eds.) (2002). *Converging technologies—For improving human performance*. http://www.wtec.org/ConvergingTechnologies/Report/nbic-complete-screen.pdf.

Schmidt, J. C. (2004). Unbounded technologies: Working through the technological reductionism of nanotechnology. In D. Baird, A. Nordmann, & J. Schummer (Eds.), *Discovering and the nanoscale* (pp. 42–44). Amsterdam: IOS Press.

Staley, T. W. (2008). The coding of technical images of nanospace: Analogy, disanalogy, and the asymmetry of worlds. *Techné, 12*(1), 16.

Tanaka, M. (2005). Toward a proposed ontology for nanoscience. http://www.cais-acsi.ca/proceedings/2005/tanaka_2005.pdf.

Chapter 12
Technology, Innovation and Life World: A Phenomenological Analysis

Dazhou Wang

Abstract It is presented that a basic strategy in philosophical analysis of technological innovation is to take the life world as the logic starting point, to focus on the tensions between the routine and innovative practices, and to explain why humans can conduct innovations based on traditions. Ontologically, technological innovations motivate and lie in the existential cycle of the human being. It is argued that the technological innovation stems from the rupture of the life world and the end of circular flow economy in face of the problematic situation and it is the process of trading, disputing, and resolving of such disputes. In this process, through a series of "mangles" and "experiments", the technological problem is constructed and then solved, and eventually a new association of human and nonhuman is established, thus the life world is renewed and even reconstructed. The essence of human existence lies in technology (being) as well as in innovation (becoming), so it is absurd to deny technology and innovation and the only choice is to do technological innovations responsibly. Responsible innovation means not so much letting responsible people into the innovation process to supervise the irresponsible innovators, but rather to help broaden the innovation vision and to share such responsibilities.

Keywords Technology · Technological innovation · Life world · Routine practice · Innovative practice · Phenomenological analysis

12.1 Introduction

The economist Joseph Schumpeter is the first person who took "innovation" as an academic concept and accordingly developed his own economic theory. In his view, an innovation includes five aspects: the introduction of a new product, the

D. Wang (✉)
School of Humanities, University of Chinese Academy of Sciences, No.19 (A), Yuquan Road, Shijingshan District, Beijing 100049, China
e-mail: dzwang@ucas.edu.cn

Q. Wang (ed.), *Chinese Philosophy of Technology*, Philosophy of Engineering and Technology 34, https://doi.org/10.1007/978-981-15-1952-9_12

introduction of a new method of production, the opening up of a new market, the conquest of a new source of raw materials, and the establishment of a new organization of enterprise (Schumpeter, 1934). Among them, the introduction of a new product and a new method of production have been collectively renamed as "technological innovation". According to Schumpeter, innovations are accompanied by the destruction of the old rules and the old order, and thus can be regarded as the process of "creative destruction".

Since 1950s, the activities of technological innovations have drawn more and more attention of both academics and policy makers in many countries. Given that an innovation is inherently an economic phenomenon, it is no surprise that philosophers have paid little attention to such a phenomenon, although innovations have long since become the main driving force for economic and social development. Since the mid-1990s, however, a few of philosophers and ethicists of technology have begun to study the social, epistemological and ethical issues in innovations or to conduct philosophical studies of innovations in order to help harness the development of technology for the welfare of mankind (e.g., Steiner, 1995; Wang, 2001; Xia & Chen, 2001). In fact, the empirical turn in philosophy of technology that emerged in the late 1990s, to some extent, means that philosophers began to pay attention to the internal process of the innovations (Achterhuis, 2001; Kroes & Meijers, 2001). At present, most philosophical studies on innovations are still mainly carried out from the perspective of ethics or epistemology (e.g., Blok, 2018; Martin, 2018), and there is a lack of philosophical investigations at the ontological level. A notable example is the study and practice of "responsible research and innovation" (RRI) originated from Europe and then spread to the whole world. The underlying idea of the RRI is that, by considering the social, ethical and legal issues in innovation, innovations will be conducted in a way that are not only economically profitable but also more environmentally sustainable, socially desirable and ethically acceptable (Schomberg, 2013; Grunwald, 2014). Such endeavors should certainly be encouraged in consideration of so many challenges raised by emerging technologies.

Different from this kind of research, this paper tries to explore the philosophical concerns of technological innovations at the ontological level with the help of phenomenology. The author argues that a technological innovation is actually a strategic site to understand the essence of human and society, and therefore it is an essential object for philosophical inquiry. It is innovations that break the black-box of human existence, disclose the reality of human existential foundation, and even renew and help reconstruct such a foundation. In other words, innovations are concerned about human's self-understanding and habitat status in the life world. Moreover, the author would like to make it clear that current philosophical studies are not enough to reveal the nature of innovations and thus to step further to present an ontologically phenomenological analysis of innovations based on the insight of phenomenologists. In so doing, the aim is to bridge the phenomenology and innovations and to clarify the intrinsic connections among technologies, innovations and the life world. The author hopes this study would be helpful for deepening the understanding of innovative practices and shedding new light on the discourses concerning the technological development and responsible innovations.

12.2 Logic Starting Point of Technological Innovation: The Life World

Realized that humans had already lived in this social, historical, political, and cultural world prior to any scientific or philosophical reflective analysis, Edmund Husserl (1954) for the first time conceptualized the totality of this everyday life and the world as life-world. As a totality what is self-evident or given, the life-world can be thought of as the horizon of all our experiences, in the sense that it is that background on which all things appear as themselves meaningfully. Life-world is more or less the background environment of competences, practices, and attitudes represented in terms of one's cognitive horizon. It is the lived realm of informal, culturally-grounded understandings and mutual accommodations (Habermas, 1987). Naturally, technological innovations can only happen in such a life world.

To understand the genesis of technological innovations, researchers need to explore human practices in general. Basically, there are two kinds of practices in human society: routine and innovative practices. The former emphasizes the habitual and rule-governed features of practices and is therefore closely linked to the concept of the life world. The latter is characterized by the break of rules and traditional knowledge bases, thereby bringing about social changes and even renewing the life world. Certainly, technological innovations belong to the innovative practices. Generally speaking, the routine practices have drawn much more attention from social theorists than the innovative practices.

Indeed, the current mainstream discussions on social practices mainly focus on the routine practices, such as those in phenomenological sociology, ethnomethodology, and so on. These endeavors have emphasized the human reflective behaviors in the life world, and tried to subvert the traditional approach of theory-dominance rooted in the Cartesian philosophy. Be it Heidegger's concept of the ready-to-hand (Heidegger, 1962), Ludwig Wittgenstein's "ungrounded way of acting" (Wittgenstein, 1979) or Pierre Bourdieu's concept of the habitus (Bourdieu, 1974), their common aim is to reveal the essence of the routine or habitual practices. As a "handedness" state, the routine practices are subject to "habituation" and "structure", so that they become a naturalized and rigidified existence with collective unconsciousness. It is true that Heidegger also discussed the "present-at-hand" state, the destruction of the "ready-to-hand" state, which could lead to further "thematic investigation", a very beginning of the innovative practices. In this way, his analysis certainly suggests the difference between routine and innovative practices. However, his main focus was on the structure of the routine practices in the life world, and the very essence of innovative practices was largely put aside.

Interestingly, the conception of the life world is parallel to Schumpeter's description of "circular flow" in his discussion of the economic system, which can be characterized as a static economy in which no innovations occur, and demand and supply are completely in accordance. It is the innovations brought about by entrepreneurs that can subvert the circular flow and make the economy evolve. Accordingly, Shionoya (2010) even proposed the "Heidegger = Schumpeter Thesis" in the economically ontological sense.

So, what state are the technology in the life world and an economy in circular flow? To answer this question, it should be firstly noted that society itself is an association of humans and non-humans, rather than a collection of pure relations among human beings (Latour, 1993). In order to further understand human practices, it is helpful to divide the non-human elements into three kinds: the natural things, institutional rules and technical artifacts. In this sense, human practices are comprised of four elements: the human, the natural, the institutional and the technological. If the social practices are governed by "habitus" and "structure", there will be no subsequent innovations, but only the inheritance and reign of traditions. In other words, the association of the human, the natural, the institutional, and the technological is in a so harmonized state that everything is black-boxed and taken for granted so that innovations could have no opportunities to occur. The traditions, including the technical ones, are embodied in the individual's body and mind, and can be inherited from one generation to another by mentoring or schooling. At the same time, they are embodied in the specific social-technical system and spreads with their reproduction. If a society is strictly adhering to all sorts of traditions including the technical ones, the economy would only be in the circulation flow, and all social practices would belong to the routine ones.

Of course, such a static state is just an "ideal type", similar to Thomas Hobbes' concept of the "jungle society", and John B. Rawls' concept of "veil of ignorance". Thus it should be considered as the logic starting point and frame of reference for the theoretical analysis of innovations. As a matter of fact, in whatever scoiety, there is always emergence of some innovations, albeit the degree varies. The occurrence of a technological innovation means breaking through the harmonized state of the life world, reconstructing the associations of humans, the natural, the institutional and the technological, and subsequently forming new routines and traditions. So, such questions as how innovations may occur and develop, and what kinds of social conditions are needed, should be analyzed in depth. In the words of Deleuze (1994), it is necessary to think about the tension between difference and repetition. Indeed, to inherit tradition means repetition, but this repetition is not absolute but a differentiated one. The key question is how to explain the generation of the differences and innovations.

12.3 Genesis of Technological Innovation: Problem Definition

It is conceivable that in the life world and circular flow economy there are lots of "mediocrities" or Das Man. According to Heidegger, Dasein as being-in-the-world is under the restraint of history, tradition, and custom and has acquired certain experience and knowledge about the self, others, and the world. Insofar as the past experiences are considered to persist, the conventional mode of action will more or less be maintained today. This is how Dasein conceives and practices his mode of action under thrownness into the world. Therefore, the unenterprising, unexamined, leisurely and carefree mood, and ubiquitous gossip about various matter of family and neighborhood, are their typical state of existence.

In fact, Dasein is always thrown into certain structures which mean the rules and resources that are referred to repeatedly in the process of social reproduction. These rules and resources intertwined in the daily life keep and sustain the social and system integration, resulting in routines that are taken for granted, and then promoting an awareness of practices to monitor their own behaviors reflectively (Giddens, 1986). In such a state, however, the survival challenges would come when they are least expected, such as natural disasters, diseases, plagues, engineering disasters, foreign invasions, and so on, which would disturb the peace of the comfortable life, break the "handedness" state of the world, and inspire Dasein to reflect on their very existence, especially on various rituals of daily life, routines and even social structures. It is in this problematic situation that innovations could emerge and develop.

Although the external stresses are the initiator of innovations, the internal "will" is even more important. Actually, while Dasein is constrained by a number of factors occurring in the past and is obliged to stay under restrictions characterized as thrownness, he can still open up a new vision of the future and create an active relationship to the world by the projection of his own possibility.

Indeed, human existence is a fateful rupture in the fabric of the world. According to Heidegger, human is a becoming, not as a maturing fruit, but as a perpetually incomplete yet always already its end, already it is not-yet. Humans can anticipate, can project future themselves, and run ahead of themselves, only on the basis of the already-there of their inherited past (Harman, 2007, 52). The essence of life lies in the transcendence which means that life projects its own higher possibility before itself, and brings itself in advance into something that has not yet been attained. More generally speaking, the being as being exists as the standing-reserve (Bestand), and at the same time exists in the continuous creation and destruction (Heidegger, 1984).

According to Friedrich Nietzsche, Heidegger and Schumpeter (Reinert & Reinert, 2006), there are two types of humans (average man or superman), or two types of existential states (authenticity or unauthenticity) in the real society. Innovators as superman exhibit much more distinctively the intellectual merit of being able to creatively solve practical and theoretical problems than average people do. It is this existential state that is called authenticity by Heidegger. The term authenticity refers to both an unconventional approach to the world and the openness to alternative viewpoints. The key moments of innovation arise when practitioners assert their individuality, break free from the structured thinking of their traditional methods, and synthesize novel interpretations that move their actions in new directions (Steiner, 1995). If one day people begin to question such an ontological basis of their existence and realize that solidification itself would bring problems, they will be motivated to innovate. When habitual tasks and procedures rupture, there will appear a "gap" between the subject and the object, and their "imperfection" will be exposed. It can be said that the gap and interaction between imperfect objects and imperfect subjects are the internal dynamics of innovative practices. The power of incomplete objects is that they provide a direction for further possible exploration. Such incompleteness generates meaning, creates practices, and facilitates the constructive extension of practices (Schatzki, Knorr-Cetina, & von Savigny, 2001). The previous practice was no longer a black box not to be questioned, but rather the subject of thematic investigations to create new knowledge with a prudent stance. Historically speaking, as an inevitable result of long-term evolution, practical activities begin to become knowledge centered and rely on experts more and more heavily. Currently, a variety of research practices permeate all aspects of social life so that people have found themselves living in the so-called "risk society" (Beck, 1992) or "experimental society" (Krohn & Weyer, 1994).

However, problematic situations would not always be interpreted as the technical problems to be solved by technical means. Especially in ancient times, the predicaments faced by the society were often explained as those of witchcraft or of religion. Many others were explained as political problems, so much so that they were solved by burning books, destroying land, restitution, royal marriage, treaties, etc. In many cases, humans even simply "naturalized" these problems as fate they had to accept. Therefore, the precondition of technological innovation is that people are willing and able to translate the difficult situations into "technical problems" and accordingly solve them by means of technological innovations. In fact, one of the basic characteristics of modern society is that people tend to think from the technological perspective no matter what difficult situations they encounter, and tend to define and solve them as technical problems. Such a tendency finally forms a positive feedback process in that using technical means to solve a problem successfully would increase the innovators' expectations, and subsequently lead to more technical problems defined and subsequently solved by technical means (Wang & Guan, 2004). But obviously this situation is not straightforward.

This actually involves the social and cultural conditions of technological innovation. The system of incentives will contribute to the development of good traits for innovations by curtailing external threats and hindering their emergence. From the historical and logical point of views, religious decline, social diversity, and market economy have played an indispensable role in propelling humans boldly onto the road of technological innovation. It is in such a social and cultural atmosphere that the fantastical entrepreneurs with "extraterritorial thinking" and strong "will to power" can emerge and flourish so as to break the routines, go beyond the secular life, and resolutely implement innovations. This kind of cultural-social soil is the very foundation of technological innovations.

This is proved by the "road to rich in the West" explored by Rosenberg and Birdzell (1987). In their view, the sustained economic growth of the West since modern times began with the emergence of a highly autonomous economic field that was not controlled by politics or religion. This autonomous economic field originated in the late Medieval Ages, when secular and religious regimes gradually relaxed or lost control over all fields of society. By the mid-19th century, the Western society had granted considerable autonomy to enterprises, of which the four autonomous rights were of special importance: individuals had the right to do business with lesser and lesser political constraints; the enterprise had the right to acquire goods and store them for resale; the enterprise had the right to increase its business and to move from one industry to another; corporate property could not be unreasonably possessed or deprived by political authorities. Such social conditions created a huge innovation passion, and enabled the West to go to prosperity through continuous innovations. In fact, there is a similar logic in China's structural transformation in modern times.

It is true that technological innovations begin with technical problems, and the so-called technical problems are those that need to be solved by technological means. In many cases, for people facing the same predicament, whether it is defined as technical or non-technical problems—such as political, social, psychological problems and so on, cannot be priori decided. From the existing predicament to the specific problem, a process of "translation" is needed, which is greatly influenced by the social and technological conditions of the time. The same predicament, in different eras, for different people, may be translated into different types of problems, such as religious, political, management or technical ones, and so on, for which politicians, entrepreneurs, managers, engineers, customers and other stakeholders would come to negotiate and struggle. Since the same predicament can be translated into different types of problems at the same time, there are different solutions. These alternatives, perhaps complementary ones, together constitute the way of getting out of the difficult situation. Technical problems have six sources (Wang, 2001): those resulted from the existing engineering dysfunction; those arose from the extension and upgrade of existing projects; those caused by the mismatch in the certain social-technical system; those caused by imbalance between competitive social-technical systems; those arose from the new development possibilities inspired by newly scientific discoveries, and those arose directly from new social needs and environmental stresses. All of these can be attributed to the mismatch within and between social-technical systems, and to the conflict between those systems and the external environment. Being driven by the problematic situation, humans would begin to conduct thematic investigations on current technologies. Such investigations would infiltrate into each element of human practices, the natural, the human, the institutional, and the technologies. From this point of view, natural science, human science, social science and technological science all have their ontologically practical bases. Once the results of the investigations are applied to human practices, technological innovations will be developed substantially.

12.4 Development of Technological Innovation: Disputes, Trading and Experiment

Technological innovations are essentially a process of mutual adjustment among the human, the nature, the institution and the technology. In this process, the "predicament" and "problem" faced by human beings are just like the light shed on the life world, which makes the subject and object emerge from "chaos". In other words, an object for a subject and a subject for an object are created simultaneously along with the innovative practice. The subject here can be understood as a practitioner community, while the object can be understood as a technical system and its products. Then, as a social process in which the subject and object create each other, how does a technological innovation unfold?

12.4.1 Disputes in Face of Technical Problems

The Innovation is a non-linear and multi-layered process which involves numerous agents and stakeholders (Schomberg, 2013). From the perspective of phenomenology, people encounter the world with a concerned attitude and are bound to situations, but from these situations they project themselves forward. Because different people have different ways of encountering the world historically, their ways of projecting themselves must be different. That being the case, different stakeholders in the process of a technological innovation usually have different perception and interpretation of the "incomplete" object, thus have different points of view about how to transform it. In this sense, a technological innovation is a process of resolving disputes between different participants. Such disputes can be divided into two categories: One is the interest conflict among various stakeholders in the process of innovation; the other is the technical disputes among engineers. The former mainly involves the need of integrating the technical and social factors, and the latter involves the need of integrating various technical factors.

The first kind of disputes is obvious to understand. Stakeholders refer to individuals, groups or organizations who are affected by an innovation project and at the same time can affect the project, such as the government, implementing agencies, the target group, or community organizations and related groups. As these stakeholders have their own perspectives and interests, and the technology innovation that is about to unfold is bound to bring them very different impacts, it is inevitable that there will be disputes about whether or not to make innovations as well as how to make innovations. Therefore, it becomes very important to conduct dynamic evaluation and audit of stakeholders in the process of innovation. For which it should be kept asking: what expectations do stakeholders have for technological innovations? What benefits or negative effects can technological innovations bring? What resources do the stakeholders have and are they willing to use them to support the project? Only in this way can innovators coordinate the relationship of interests and lay a foundation for the development of technological innovations.

The second kind of disputes is much more invisible than the first one. In the process of innovation, technical personnel usually come from different fields, adhere to the very different ways of thinking and behavior standards, have different technical abilities and experiences, use very different instruments and equipment, and speak very different professional languages, therefore it is very difficult for them to communicate and make compromises. It can even be said that technical personnel from different fields live in incommensurability object worlds. As a result, they often have different views on technology innovations, especially on the relative importance of key design parameters (Bucciarelli, 2003). In this case, disputes among them are inevitable. To tackle such disputes, one way might be to sit down and try to break the task down into a set of sub-tasks, and then engineers work on their own. Quite understandably, it should not be hard for routine practices, as experienced teams have already been familiar with them. For innovative practices, however, this is often unlikely. It is much uncertain to decompose the task and to set boundaries, for nobody can fully understand the overall tasks in advance. In this case, all of the participants need to be interactive with exploratory spirit, and thus coordination is always a challenge. Of course, that engineers in different fields live in different object worlds doesn't mean they have nothing in common. After all, they still speak everyday language, share some ontological beliefs about how the world works, have some common understandings of each other's technical field, and have some trust in the basic principles from different disciplines. All of these can help resolve their disputes and finally achieve agreements.

To go a step further, there are some interactions between the two kinds of disputes. The settlement of the first kind of disputes usually comes through the settlement of the second kind of disputes with technological means. And if the second kind of disputes could not finally be resolved, it would return to the first type of disputes and be dealt with by social means. In general, only when the conflicts of perspectives and of interests are dissolved in the specific engineering design, and the interests of all parties are balanced through innovative design to a certain extent, can the disputes be resolved satisfactorily. The Three Gorges Project, for example, has solved the migration of the Chinese sturgeon through specific design methods, which was a particular concern for conservationists. It should be noted that disputes are not so much a negative phenomenon to be overcome, but rather a phenomenon that stimulates innovation and gives it a sense of direction, thus having positive value that cannot be ignored.

12.4.2 Trading Zone in Technological Innovation

To further understand the resolution of disputes in technological innovations, the concept of "trading zone" is needed. It is in the trading zone that various parties can settle their disputes, and thus promote technological innovation to succeed. The trading zone is a communication platform and negotiate space, either geospatial or virtual, facilitating the settlement of coordination problems in technological innovations. In this space, the stakeholders have opportunities to achieve "harmony without uniformity".

Peter Galison is the first historian of science to use the trading zone concept inspired by anthropology studies to explain how physicists in different paradigms can work together and even cooperate with engineers to develop particle detectors and radars. According to him, the two groups could reach agreements on the rules of trading, even if they gave completely different meanings to the item being traded, for the trading partners could work out a local reconciliation regardless of their overall differences. To him, interactions among all parties may inspire delicate language for communication, from the most extreme jargons to half specialized Pidgin, and then to full-fledged Creole capable of supporting a variety of complex activities (Galison, 1997). For example, in the case of the development of radar, physicists and engineers invented such a shared concept as "equivalent circuit" which physicists understood as a symbolic expression based on the field theory and which engineers regarded as an extension of the radar toolbox. Of course, trading can also be mediated by certain agents who are familiar enough with the languages of two or more cultures. The ability to communicate skillfully in more than one discipline is called interactive expertise (Gorman, 2010). Anyone who wants to accumulate interactive expertise must be immersed in his own field and other areas of expertise for several years.

According to the theory of trading zones, technological innovation can be regarded as the superposition of several trading zones. One of the major tasks of the innovators is to identify the trading zone around the technological innovation, and then to facilitate dialogues and transactions among various stakeholders. In general, there are three ways to do transactions: The first is to gradually create a discourse system, from the most extreme jargons to the half specialized Pidgin, and then to the Creole, to support dialogue and exchanges; The second is to rely on specific agents with interactive expertise as mediums to help enable transactions; The third is to resort to specific boundary objects to simplify transactions. These three modes of trading may be conducted in parallel or separately. These transactions include not only those between human actors, but also the property exchanges between human actors and objects, and between different objects. The concept of trade zone highlights the heterogeneity and complexity of innovation practices. It is through such transactions that technological problems are likely to emerge and be finally resolved.

12.4.3 Mangles and Experiments in Technological Innovation

It is conceivable that trading in a trading zone cannot be once and for all, and even the trading zone itself changes with technological innovation. After all, innovation means uncertainty, and unintended consequences are always inevitable. Fundamentally, innovators "inform" a new possibility of human existence, and thus play a role of destroyers of anything rigid. To destroy these rigidities and realize such new possibilities, they have to test beings and bring them into the experimental process. Basically, a technological innovation is an experimental process with mangles of various forces and the emergence of new things. In this process, the innovators gradually increase their knowledge, feel the resistance of obstacles, expand the scope of control, and even adjust their targets at the same time, so as to achieve a satisfactory result. According to Andrew Pickering (1995), the mangle is a temporal emergent process, and its results cannot be given in advance, which means that attention should be primarily paid to the post-humanist view of the interwoven and constructional relationship between human and non-human, and to the fact that there is nothing constant in the innovation process. Actually, the traditional social theories have always tried to abstract a fixed framework from the flowing phenomena, while the emergent phenomena self is contrary to such a claim. This requires research focusing on the places where human beings meet material forces, such as laboratories, factories, battlefields, and so on. From this point of view, research activities actually are rooted in human practices, continuously growing from the encountering places of material power and human power, and finally returning to such sites.

In fact, in the process of technological innovations, the material, conceptual and social forces co-evolve. The establishment of new products and production procedures, and the formation of new knowledge systems and social arrangements are firmly linked, reinforcing, constructing and binding each other, in which there are full of mangles and experiments with the emergence of new things, products, technology, organizational forms, social relations, and so on. It is because of the inherent limitations of human knowledge that no one can completely control the innovation, and that innovators have to face the emergence of new problems and things over time.

A Technological innovation includes both incremental and revolutionary changes, and the mangle and experiment involved in the latter case would be more austere. Many technological innovations are realized through gradual accumulation and improvement. By contrast, the breakthroughs have potential to break the old structure and to form a new one, and thus will have major impacts on economic and social development, even reconstructing humans' perception of space and time. Such innovations often involve the transformation of technological paradigms. Certainly, the new paradigm is not necessarily able to seize the power smoothly in competing with old one. This is due to the inertial power of the traditions and structures. Traditions have solidified in the existing structures, and embodied in habits, strategies, organizations, user networks, supply networks and so on, by which innovations tend to be locked in certain trajectories, showing huge "organizational inertia" (Hannan & Freeman, 1984), which is more conducive to the old paradigm. Such competition would also cause the old paradigm to change itself and even realize the huge leap within the paradigm. The failed new paradigm does not necessarily disappear, rather, it is also possible to find its own existential space, and may have opportunities to come back in the future.

12.5 Accomplishment of Technological Innovation: Return to the Life World

With the closure of the disputes, mangles and experiments, a technological innovation will achieve success, which means the new technology has been embedded in the life-world and the new harmonious state among the natural, the human, the institutional and the technological factors are finally be realized. Ontologically speaking, the existential cycle of the human being has evolved from the life-world to the renewed life-world with new technologies. Of course, in this process, the "routines" and "traditions" which are fixed and govern the life-world, are impossible to be completely abandoned On the contrary, they have positive values for innovation in many cases. Just as Heidegger argued, "Creation needs what is fixed, first, in order to overcome it, and second, in order to have something that has yet to be fixated, something that enables the creative to advance beyond itself and be transfigured" (Heidegger, 1984).

It is in the process of inheriting traditions and at the same time renewing them to some extent that technological innovation can be successful. As a process in which subjects and objects mangle and create with each other, a technological innovation certainly results in major effects. First, with the emergence of a technological innovation, such a new collective of actors have been created at the same time. In this collective, there are new customers, new suppliers, new engineers, new workers, new investors, even the new public. It is these new agents who have realized the potential power of technological innovation. Thus, technological innovation is the effective way to cultivate new agents. It is through the practice of training and mentoring by predecessors, a new generation of skilled technical personnel and engineers as main inheritors of engineering traditions can be forged. Second, the technological innovation can bring a series of subsequent innovations, which may lead to the emergence of innovation clusters. The combination of competition mechanism, learning mechanism and chain mechanism will inevitably bring innovators competing with each other and the swarms of heterogeneous innovations. With the maturity of technologies, the new technology tradition will be formed finally, companied by the organizational inertia and even the unauthenticity state of Dasein. Third, technological innovation will change the social life. As technology innovations evolve, new knowledge and technologies, new artificial systems, and new modes of institutions and organizations will emerge. These new things will challenge the life-world at the beginning, and eventually open up a new living space, new living possibilities, even new languages and new living consciousness. All of this means that technological innovation is at the root of social change.

It is made clear that the technological innovation is actual an embedding as well as mangling process of new technologies into the life-world, through which technologies are black-boxed and integrate in the new technical tradition, while people tend to forget these technologies as the foundation of their existence. Just as Alfred North Whitehead said, "It is a profoundly erroneous truism, that we should cultivate the habit of thinking what we are doing. The precise opposite is the case. Civilization advances by extending the number of important operations which we can perform without thinking of them" (Whitehead, 1911). This is a process of "phenomenological transformation". To achieve technological innovations, one must firstly conduct thematic investigation into the object as well the life world by which he could acquire much new knowledge, especially those articulated propositional and procedural knowledge. However, once such technological innovations are accomplished, all needed knowledge must be further reintegrated into practices and made to work invisibly, implicitly and unconsciously. Only then can this knowledge exert its best functions. This is why Michael Polanyi claims that the goal of "proficiency" is achieved by following a set of rules that the operator himself does not understand (Polanyi, 1958). In other words, "unconsciousness" is also a necessary condition for proficiency, and that trying to pay attention to details could often be destructive.

Only at this moment, four typical relations, proposed by Ihde (1990), among human, artifacts and the world would appear: embodiment relation, hermeneutic relation, alterity relation and background relation. The central questions in Ihde's research are what role technology plays in everyday human experience and how technological artifacts affect people's existence and what relations technologies have with the world. However, the four relations are those between the "given" user and the "given" artifact under the "given" life world scenarios, excluding the running in process and dynamic interactions among the users, artifacts and the world, and thus are only four special cases in the state space of possible relationships (Zhang & Wang, 2015). With this lens, technological innovation always necessitates the artificial manipulation of the possible relations among human, artifacts and the world: the enlargement or reduction, concealment or disclosure, and ultimately the formation of a stable or multi-stable human, artifact and world relationship.

In short, the process of technological innovation means a circle of phenomenological transformation. That is to say, technological innovation starts from the rupture in the life world. After a series of agitation, it finally returns to the life world and renews the life world by a process of black-boxing. This reversion is not simply an embedding process but will be done through mangles and experiments.

In fact, the life world is not constant, but can be reconstructed through "re-structuration", which includes both "repetition" (reproduction) and "difference" (renewal). This renewed life world will be the new starting point for the subsequent technological innovations.

12.6 Conclusion

If technology is the root of human existence, technological innovation, based on technology and related traditions as the ready-made things, is the explorative journey to create new possibilities for human existence. Technological innovations mean breaking the harmonious cage of the life world, rebuilding the associations of humans, the natural, the institutional and the technological, and helping form new routines and new traditions. A basic strategy in philosophical analysis of technological innovation is to take the life world and "circular flow" economy as the logic starting point, to focus on the tensions between the routine and innovative practices, to think about the genesis and development of innovations, and to explain why humans can conduct innovations based on traditions.

Ontologically speaking, technological innovations motivate and lie in the existential cycle of the human being. Technological innovations stem from the rupture of the life world and end of the "circular economy". With such ruptures and ends, all the basic elements of the social practices, i.e. the human, the natural, the institutional and the technological, would be taken out of black box and brought to light. The technological innovation can be regarded as a trading zone, and the innovation process is the process of transactions among various stakeholders who dispute and try to resolve these disputes. In this process, through a series of mangles and experiments, technological problems will be firstly defined, then gradually be resolved, and eventually form a new collection of human and nonhuman and structure new routines and new time-space areas, that is to say, a renewed harmony of the life world. So, in the context of technological innovation, the human, the natural, the institutional and the technological are in a state of continuous flow, and therefore we can no longer simply use the binary thinking modes such as "natural and artificial" and "technology and society" to elucidate technological innovations.

Indeed, technology does not equal to innovation from either economic or ontological perspectives. A static society has only solidified technologies but technological innovations. Adhering to harmonious state is not always a good choice. Instead, it would block the path of human being as being. The essence of human existence lies in technology (being) as well as in innovation (becoming), especially technological innovations. To break the solidification of a society, the fundamental means is to conduct technological innovations. So it is absurd to deny technologies and innovations; instead, the only choice is to make technological innovations responsibly. In doing so, it requires to develop the supporting social institutions to release Dasein's immanent innovative spirit. However, technological innovation is indeed a risk, and so a strong sense of responsibility is required.

Ontologically and historically, to some extent innovators has always been responsible, otherwise they would perish very long ago. Such being the case, responsible innovation means not so much letting responsible people into the innovation process to supervise the irresponsible innovators, but rather helping broaden the innovation vision and to share such increasingly weighty responsibilities. This is necessary because today's innovations are so complicated and influential that they cannot be left to innovators to make their own decisions.

References

Achterhuis, H. (Ed.). (2001). *American philosophy of technology: The empirical turn*. Indiana: Indiana University Press.

Beck, U. (1992). *Risk society: Towards a new modernity*. London: Sage.

Blok, V. (2018). Philosophy of innovation: A research agenda. *Philosophy of Management, 17*(1), 1–5.

Bourdieu, P. (1974). *Outline of a theory of practice*. Cambridge: Cambridge University Press.

Bucciarelli, L. L. (2003). *Engineering philosophy*, Delft University Press, 11.

Deleuze, G. (1994). *Difference and repetition*. New York: Columbia University Press.

Galison, P. (1997). *Image and logic: A material culture of microphysics* (p. 783). Chicago, Illinois: University of Chicago Press. 783.

Giddens, A. (1986). *The constitution of society: Outline of the theory of structuration*. Cambridge: Polity Press.

Gorman, M. E. (Ed.). (2010). *Trading zones and interactional expertise*. Cambridge, MA: The MIT Press.

Grunwald, Armin. (2014). The hermeneutic side of responsible research and innovation. *Journal of Responsible Innovation, 1*(3), 274–291.

Habermas, J. (1987). *Theory of communicative action, volume two: Lifeworld and system: A critique of functionalist reason*. Translated by A. Thomas. McCarthy, Boston, MA: Beacon Press.

Hannan, M. T., & Freeman, J. (1984). Structural inertia and organizational change. *American Sociological Review, 49*(2), 149–164.

Harman, G. (2007). *Heidegger explained: From phenomenon to thing*. Chicago, IL: Open Court.

Heidegger, M. (1962). *Being and time*. London: Harper & Row.

Heidegger, M. (1984). *Nietzsche*. Translation by David Ferrell Krell. San Francisco: Harper and Row Publishing.

Husserl, E. (1954). *The crisis of European sciences and transcendental phenomenology*. Evanston: Northwestern University Press.

Ihde, D. (1990). *Technology and the lifeworld: From garden to earth*. Bloomington: Indiana University Press.

Kroes, P., & Meijers, A. (Eds.). (2001). *The empirical turn in the philosophy of technology*. Amsterdam: Elsevier Science.

Krohn, W., & Weyer, J. (1994). Society as a laboratory: The social risks of experimental research. *Science and Public Policy, 21*(3), 173–183.

Latour, B. (1993). *We have never been modern*. Cambridge, MA: Harvard University Press.

Martin, S. (2018). The virtues and vices of innovators. *Philosophy of Management, 17*(1), 79–95.

Pickering, A. (1995). *The mangle of practice: Time, agency, and science*. Chicago: University of Chicago Press.

Polanyi, M. (1958). *Personal knowledge: Towards a post-critical philosophy*. Chicago: The University of Chicago Press.

Reinert, H., & Reinert, E. S. (2006). Creative destruction in economics: Nietzsche, sombart, schumpeter. In J. G. Backhaus, W. Drechsler (Eds.), *Friedrich Nietzsche (1844–1900): Economy and society* (pp. 55–85). Springer.
Rosenberg, N., & Birdzell, L. E. (1987). *How the West grew rich: The economic transformation of the industrial world*. New York: Basic Books.
Schatzki, T. R., Knorr-Cetina, K., & von Savigny, E. (Eds.). (2001). *The practice turn in contemporary theory*. London: Routledge.
Schumpeter, J. A. (1934). *The theory of economic development*. Cambridge, MA: Harvard University Press.
Shionoya, Y. (2010). Hermeneutics and the heidegger = schumpeter theses. *American Journal of Economics and Sociology, 69*(1), 188–202.
Steiner, C. J. (1995). A philosophy for innovation: The role of unconventional individuals in innovation success. *Journal of Product Innovation Management, 12*(5), 431–440.
von Schomberg, R. (2013). A vision of responsible innovation. In R. Owen, J. R. Bessant & Heintz M (Eds.), *Responsible innovation: Managing the responsible emergence of science and innovation in society* (pp. 51–74). Wiley.
Wang, D. (2001). *Technological innovation and institutional structures*. Shenyang: Northeast University Press.
Wang, D., & Guan, S. (2004). On philosophy of technology, technological practice and rationality. *Philosophical Researches,* (11), 55–60.
Whitehead, A. N. (1911). *An introduction to mathematics*. London: Williams and Norgate.
Wittgenstein, L. (1979). *On certainty*. Oxford: Basil Blackwell, Verse 110.
Xia, B., & Chen, C. (2001). On the philosophical study of technological innovation. *Studies in Dialectics of Nature, 17*(8), 18–21, 35.
Zhang, B., & Wang, D. (2015). On phenomenology of artifacts: A new analysis framework. *Philosophical Researches,* (6), 113–119.

Chapter 13
Philosophy of Information Technology and Its Contemporary Value

Feng Xiao

Abstract Information technology (IT) is the technology that determines the social characteristics of today, so the philosophy of information technology (PIT) which takes information technology as the research object is undoubtedly the most important branch of philosophy nowadays, and also the contemporary form of the philosophy of technology. The main research contents of the PIT can be constructed based on the main research fields of philosophy, so as to form the ontology of IT, epistemology of IT, and axiology of IT. Each part contains new philosophical issues and a new paradigm shift. Modern information technology is expanding the new frontier and bringing new challenges of philosophy, including the Internet of things, artificial emotion, artificial experience, brain-machine interface and so on. In this article, we take the latest development of these four areas and to explore their philosophical implication and value, thus presenting that the PIT is an open system, and a kind of philosophical research driven by frontier questions.

Keywords Information technology (IT) · Philosophy of information technology (PIT) · Paradigm shift · Frontier technology

The philosophy of information technology (PIT) is the product of the information age and it is a philosophy based upon information technology (IT). The arrival of information age was symbolized by the birth of modern IT, such as computers and the internet, which have caused great and profound changes to the whole world. This epochal transformation means that modern IT has changed the history, reshaped the society and renovated the civilization form. It is also the "engine" of human mental evolution and the direct driving force to push the philosophy of technology (PT) to the new form of PIT. Since its birth, PT has manifested its own core features and status, and formed special research contents. What runs through it is the transformation of the paradigm of philosophical research. As a new form of

F. Xiao (✉)
South China University of Technology, No. 381, Wushan Road,
Tianhe District, Guangzhou, China
e-mail: xiaof56@sina.com

Q. Wang (ed.), *Chinese Philosophy of Technology*, Philosophy of Engineering and Technology 34, https://doi.org/10.1007/978-981-15-1952-9_13

PT, the development of PIT has brought about a series of noteworthy frontiers, leading us to think more deeply about some fundamental issues of philosophy.

13.1 The Nature and Position of PIT

Information technology determines the appearance of our time, and philosophy shoulders the historical mission of focusing on the new developments of the time and the new trends in science and technology. Therefore, philosophy must lay emphasis on IT. PIT is the fruit of a "two-way" influence, the collaborative achievement of upgrading modern technological issues to philosophical ones and instilling philosophy in modern technology.

PT should reveal the deep impact of technology on the society and humanity. Today, when IT has the function of creating the features of the times, it gains the features of being the "dominant technology" of all technologies. The social and humanistic functions of technology will inevitably be transformed into the function of IT. The philosophical issues of technology are mainly reflected in those of IT. As IT is becoming the dominant technology, PT must also be transformed, and the PIT has become the contemporary form of the PT.

PIT can be considered as a specification of general PT, and the transformation of PT to PIT can be seen as a transition to a departmental PT. PIT can also be seen as a generalization and promotion of the emerging computer philosophy, digital philosophy, cyber philosophy, network philosophy, and big data philosophy. In this sense, it can be seen as a transition of PT from the specific level to the general level. And computer philosophy, digital philosophy, etc., can be considered as branches or sub-disciplines of PIT. The prosperity of the branches provides an intellectual foundation and creates a precondition for the overall PIT study. But their prosperity is by no means a substitute for that of the PIT.

Obviously, to build PIT, it is necessary to ask some questions and conduct some research concerning the general theory of the PIT. Without such research, the great philosophical significance of IT cannot be fully grasped. In view of this, it is necessary to analyze the holistic IT with a broader philosophical vision to form a general philosophical study of IT. This "totality" includes: firstly, the totality of time—not only contemporary IT, but also historical IT. Secondly, the totality of categories—not only computers, networks, etc., but also observation instruments, display means, information loading devices, etc. This totality is to extend IT to a general technological paradigm so that the object of PIT rises from specific IT to general IT, and to further extend the scattered research results of PIT into the general philosophical paradigm. PIT has also been linked to other humanities and social sciences on IT, such as the ethics of IT which can be regarded as a member of the PIT family, as it provides rich ideological resources for PIT. The PIT can be cross-linked or associated with many existing and emerging disciplines relating to information science and technology. It is leading to a new flexible integration.

PIT has its own disciplinary orientation. It can be regarded as a sub-discipline of the PT. Though the level is lower, it has great significance in that it is not only the contemporary form of PT, but also the new form or new paradigm of philosophy in the information age. It is a representative disciplinary of philosophical transformation that accompanies the transformation of the times and of civilizations.

As a special issue of subject attribute, the relationship between PIT and philosophy of information (PI) also requires attention. Both PIT and PI come from the information revolution. Their relationship is concerned about the relationship between philosophy of science and PT in the field of information science and technology. Their research objects and questions often form an intersection, such as the virtual reality, memory transplantation, mental enhancement, the relation between body and mind, and other issues. All of these are involved in both PIT and PI. In this sense, they can share some research fields and achievements. From the objective background, it is the importance of IT that leads to the importance of information, making the problem of "essence of information" as the key issue of PI. It can be said that the advent of the IT age has led to the birth of PI, which is based on the essence of information and makes us view everything from the perspective of information.

After the birth of the information philosophy that uses information as the basis for explaining everything, when exploring the nature of information and the artificial manufacturing and dissemination of information, when discussing the main activities of human beings in the information age, it is necessary to involve IT and further PIT.

While realizing the similarities and connections between PT and PIT, we should not ignore their distinctions. At present, PI at home and abroad has shown a good momentum of development. The philosophical research on information has produced abundant views and opinions. The definition of information is inconsistent in that it is difficult to reach a consensus on the issue of "what information is". The reason why there is so much ambiguity in the "information" in PI is that the philosophical distinction between "information" and "signal" is not taken seriously, so that information itself is often confused with its carriers. At a result, information is frequently confused with IT. In order to philosophically clarify the meaning of information, it is essential to clarify the meaning of IT and the connections and distinctions between information and IT in a philosophical way. To advance the study of PI, it is necessary to move moderately from PI to PIT, so that a look back of the fundamental issues in PI can be made in a broader vision.

13.2 The Main Content of PIT and Its Paradigm Conversion

The system of PIT can be constructed in many ways. Some scholars mainly build their PIT system from specific issues or interdisciplinary fields. (Mitcham, 2004) Another noticeable research approach is that can be constructed from the main field of philosophy. Every field contains new philosophical issues of technology, among which there is a new transformation of research paradigm.

1. **The ontology of information technology**

The ontology of IT includes the essential theory of IT, ontological (about Being and existence) theory of IT, and realism of IT. The essential theory of IT is to solve the problem of philosophical meaning of IT. The object of PIT is IT. Its primary task is to define the "information technology" from the philosophical perspective. The definition of IT is to answer "what information technology is", *that is*, to reveal the essence of IT. The philosophical meaning of "information technology" comes from daily usage, and yet is different from daily usage. Philosophically, it can be defined as "a type of technology that exercises information processing or symbolic conversion functions and serves people's spiritual needs", thus distinguishing it from the material production technology (Xiao, 2012).

The ontological theory of IT analyzes IT from the perspective of the philosophy of "Being". IT per se exists as a material artificial object. However, in essence or function, it creates a kind of "informational Being" which is different from the "material Being". There is a complex relationship between the two kinds of Being, which reflects the characteristics of IT and the essence of information. If it is acknowledged that all information is the product of communication among people, then all information is artificial information and is related to human. Therefore, IT has created the existence of information. In other words, information is a technical existence with a kind of dependence on IT. If IT is a kind of existence, then information is a new type of existence created by the existence of IT, so it is a kind of "secondary existence". The display and dissemination technology of information further demonstrate the role of information as a Being of "serving people". Through the "gaseous information" (the information formed from one's face-to-face communication), the "solid information" (the text information displayed by paper) and the "electrical information" (the information displayed and transferred by electronic information technology), it moves towards an increasingly enriched beings, and changes people's own existence (survival) situation. This means that the ontological consideration about IT could not be separate from "we", or the ontological issue of IT is closely related to the "we" who think about the issues of IT.

The realist research on IT discusses the realistic issues of the instruments as hard carriers of information and the symbols as soft carriers of information, as well as the realistic issue of the software system as the control system for hardware operation. These discussions have introduced new meanings into the "reality" as an important philosophical concept, changed previous views of reality, provided the reality with new forms, and realized the semantic expansion of the reality in the IT era. For example, the reality of symbol cannot be understood in an absolute sense. Without semantics, symbol is an "empty material shell" alone. Though it is also a kind of reality, it is meaningless. The reality of software also needs to be understood from a unique perspective, *that is*, the software translates information into a series of physical instructions and then implements them. When people watch the printout of a program, what they can see are lines of statements written in formal languages. But when someone holds a floppy disk with the same program in his or her hands, he or she feels the weight of the artificial product. In addition, as the objective operation state of computers is restricted by both hardware and software, software is actually a constituent element of the real operation of computers. It is one of the decisive conditions of function of computers. Thus, the reality of software is not necessarily understood from the perspective of entity reality, but rather from the perspective of activity and processing or functional reality. If the software is used as a digital symbol system, its reality resembles the "semi-reality" of symbol. Thus, information technology makes the boundaries between the real and non-real become blurred. This is especially reflected in virtual reality, virtual life, web-intercourse, and so on. IT also brings humanistic meaning to reality which is meaningless when separated from human. This brings us more hierarchical and contextual problems of reality, from which more philosophical implications of reality can be developed.

2. **The epistemology of information technology**

At the time of its origin, information technology mainly plays the role of knowing the world and exists as a "cognitive technology", so it has inherent epistemological tendencies, and thus rich in contents of epistemology. The historical development of human cognition is largely restricted and promoted by IT. Today, the involvement of computers and internet in human cognitive activities has created and is creating new characteristics of human cognition. It draws a more complex and novel picture for epistemology, and also contains a new transformation of epistemology paradigm.

The first transformation is to move from naturalized epistemology to technological epistemology. In a general sense, technology and nature are different things. Technology is artificial and thus unnatural, and nature is "non-artificial" and thus non-technological. In this sense, the "naturalized epistemology" is the cognition that the subject has when confronting natural objects by natural means, while the cognition that the subject has when confronting artificial objects by natural means. On the other hand, "technological cognition" is the cognition formed when people use technological means to know natural objects or intermediaries constructed by technology. In the era of oral communication, people's cognition was only by virtue of their own organs and the natural channel (air), and there was almost no technological device to assist or intervene in the process of cognition. The invention of characters, especially the use of printing, that opened the course of technological involvement in cognition and the types and scope of the technical intervention were expanding ceaselessly. With the full involvement of IT in the information age, the technicalizational epistemology has shown a trend toward a leading position. Nowadays, people are confronted with fewer natural objects, but more that appear on technological devices, particularly on electronic screens. Through the tool of cognition, we complete epistemic tasks more by computer, network, or AI instead of "natural intelligence" organ (human brain). In terms of cognition, more fragmental and imagery traces created by multimedia technology have appeared. In the future, along with chip implantation into human brains, memory transplantation, neurological operations, and artificial intelligence's "back-feeding" of human intelligence, human as the subject of cognition will be more enhanced by IT, which further will result in the technicalization of the subject, or form the so-called "human-machine cognitive subject". The involvement of IT will even enable our "experience" to be generated in a technical manner. It will also enable "tacit knowledge" to spread beyond language, and even allow direct communication between different minds. In short, IT is working with human to build knowledge or knowledge products. As the contemporary form of technicalizational epistemology, the epistemology with the involvement of IT fully shows the significance of the development of IT to epistemology research. IT is not only a cognitive tool, but also an internal structure in cognition process.

Therefore, what kind of IT is used to carry out the understanding depends on the corresponding "technical route" and "world construction", both of which to some extent also determines what a corresponding cognitive result to be obtained. Grasping the IT technical turning of epistemology in present age through the research for IT-involvement epistemology is the result of adapting to the era technical transformation. If this transition is not followed, the most important foundation of contemporary epistemology research will be lost. Nowadays the epistemology is no longer the "pure epistemology research" without technological involvement. The degree of attention to the IT will determine how far epistemological research can possibly go.

The second transformation is from conventional epistemology to innovative epistemology. As a powerful auxiliary tool, IT has greatly improved human cognition efficiency and understanding ability. For example, with the help of internet, human can effectively search information, disseminate knowledge, and share ideas. With the help of computer, big data, artificial intelligence and so on, many conventional cognitive activities can be referred to these auxiliary means, which liberate human from simple or low-level mental labor. There is more remaining time left for human than before, and the so-called "cognitive surplus" is formed. This is a new available ability for human cognition so as to provide more possibility for creative cognition. After breaking away from repetitive mental labor at a low level, more people turn to creative activities such as discovery, invention, and innovation which are more human-oriented. In this way, the main field of epistemology also needs to be turned from traditional epistemology to creative epistemology in order to provide more effective ways of cognition and guidance for the innovative subjects. Such an innovative epistemology can also become a new direction, which will lead to creative activities and create valuable mental results as a cognitive capacity of "cognitive surplus". This is also the view held by Clay Scheler in book *Cognitive Surplus*. Specifically, people are free to participate and share information in the network. They use free time to create more than just consume information, so that everyone becomes a potential producer of network information and provides opportunities for each other's creation. Due to many cognitive surpluses in the world, along with the aggregate value or convergence effect created by new information exchange media and communication tools, tiny achievements will accumulate into huge ones. In a word, the age of information technology is an era of "people's innovation", and the research of innovative epistemology will undoubtedly flourish.

The third transformation is from elite epistemology to mass epistemology. Knowledge is the highest achievement of cognition. The production of knowledge is undoubtedly an important cognitive activity which contains questions about how knowledge is generated, who are main producers, how knowledge production is changed in history, and so on. If we pay attention to the question of who are main producers of knowledge, we will find that they are intellectual elite for a long time. The outstanding individual figures in the name of "scientists" "thinker" "experts" occupy the "possession right" of all kinds of knowledge, from which we cannot see the trace of the common people. The work of summarizing the achievements of human knowledge, such as compiling encyclopedias, is monopolized by a few "intellectual elites". In the information age, a wide range of public participation is an important and new feature of human activities, which is also reflected in knowledge production activities. The popularity of the internet has made it a cognitive means for millions of the common people. They can spread their exploration about nature, literary creation, and philosophical meditation in the public space of network. They can also participate in the compiling of the Internet Encyclopedia and show greater superiority than the traditional paper encyclopedia compiled by a small number of elites, such as fast updating, large number of entries, high degree of sharing, free use, convenient query, and so on. Therefore, its influence and efficiency far exceeds the traditional encyclopedias. Millions of the common people as common knowledge creators and donors have contributed to human knowledge. As "volunteers", they use their spare time to contribute their own wisdom and labor without fame and fortune. Co-creation and sharing have become the "new normal". This epochal transformation of the way of knowledge generation makes the emphasis of epistemology turn from individuals to groups. This trend of epistemology has also made the collective wisdom in knowledge creation truly manifested, expanding the scope of knowledge producers and innovation subjects. The development provides a strong mass foundation for the theory of innovative epistemology mentioned above. The public epistemology of the new technology era has also put forward new topics for the study of epistemology, for example, how to integrate the activities of knowledge creation with the internet, how to make knowledge generated on the network platform reliable and authoritative, and how to overcome disorder and confusion through new knowledge management methods. The solutions of these problems will help further promote the healthy growth of internet-based mass epistemology, and create greater social and spiritual values.

The fourth is from macroscopic epistemology to meso (intermediate perspective) epistemology. Traditional epistemology is usually limited to the "grand narrative" of the cognitive process or the abstract stipulation of the essence of cognition without real and specific content of the mechanism of cognition. In the information age, with the development of IT and increasingly advanced method and the intelligent simulation technology, there are deeper and more broadening understanding for neural activity, electroencephalogram process and information symbol transform in human cognitive process. It leads to the rise of related new theories, new genres and new perspectives. It also provides fresh "blood" and "nourishment" for the philosophical research in the mechanism and essence of human cognition, greatly broadening the horizon of epistemology research. Nowadays, if we integrate the new achievements in the field of neuroscience, cognitive science, AI science, and robotics into the study of epistemology, we can greatly push it towards a new world. For example, if the concepts of sign information processing, function and computing, brain waves, neurohumoral transmission, neural network connection, pattern recognition, brain computer interface, and so on, can be integrated and refined into philosophy content or be extended to new epistemological problems, there is no doubt that we will reach a new understanding for the process of cognition. It will enrich the physiological and mental bases of reflecting objects, selecting information, and constructing knowledge. Such epistemology is no longer a macroscopic explanation of cognitive phenomena, nor is it a partial understanding which is reduced to a specific science, but a study on understanding with the "middle" characteristics. It can become a bridge for connecting the micro and macro cognition, filling the gap between epistemology and specific cognitive science research as well as realizing the reciprocity between classical epistemology and modern specific cognitive science, brain science, and intelligence science.

3. **The humanism and axiology of information technology**

Human issue is usually unavoidable in philosophy, so that philosophy is regarded as humanology in the eyes of many philosophers. Even in PT, "technology and human" is also an important part. In this sense, "IT and human" should also be an important part of PIT. It can reveal an extremely rich philosophical relationship between IT and human. It also provides new insights for understanding "the essence of human", and expands the new form of "digital development". Simultaneously, it gives us new ways of understanding human values, thus promoting the new exploration of humanology.

The contemporary development of IT is a challenge for the traditional understanding of "what human is", which affects the definition of human. The most typical one is the discussion of Cyborg. Cyborg is defined as a person's a physiological process that his/her physical energy extended by the machine to exceed the limits of human physical capacity, or assisted or controlled by a machine or an electronic device to some degree. Thus, its Chinese translation could be "电子人" (semi-electronic human), "半机械人" (semi-mechanical human), "仿生体" (the bionic body)and so on. In Cyborg, human and technology have merged, including that more devices are implanted in the human body, especially the electronic devices implanted in the human brain. The boundary between the human and the computer will become less obvious. As a combination of machinery and human, Cyborg is partly organic and partly inorganic. It means that information technology is starting to become part of our body, just like the functions of prosthetic technology, which replace or increase biological function and turn humans into Cyborg, thereby changing human nature. (Brey & Soraker, 2009) If we call the conventional standard of defining a human as a kind of "humanistic identity", then the transformation of human by contemporary IT is a big challenge to this humanistic identity. Excessive adherence to the customary human identity may hinder the scientific understanding and technical improvement of human beings, like the obstacles occurred in history. But at the same time, we must be very cautious when we use technical means to transform human beings thereby making breakthroughs in traditional humanistic identity, because not every breakthrough is positive or has a humanistic value. This is especially the case when this kind of transformation concerns "what human is" and the standards of humanistic identity.

The change of human character under the involvement of IT can be called "the digital development of human beings". It has existed objectively since the invention of computers. It can be divided into two aspects, i.e. vitro digital development and vivo digital development of human. The former case is that the mental content or thinking (as information) is copied to an outside machine (it can be called "mind-cloning") and improved according to our needs. The latter is implanting chip into the human brain. Of course, there is also a mixture of both, such as the transplantation of memory between different people. Or it can also be divided into digital therapy and digital enhancement. The former is used to change the abnormal mental state, while the latter is used to obtain better mental quality based on normal condition, and even to design "super excellent" personality that cannot be possessed by common people. This process is called the "digital enhancement".

If this kind of digital development is realized, it would trigger some very sharp humanistic problems. The most prominent of them is the relationship between human body and mind. From the perspective of PIT, the development of human body and mind are that of human information form and its carrier form. The human body is the carrier of human memory (or human information form, including self-consciousness, knowledge, experiences and so on). The human development with the involvement and help of IT is nothing more than the development of two aspects above. For example, memory transplantation is the development of information directed to people, while transplanting human memory to supercomputer is the development of the carrier. The accompanying questions are that whether the new carrier can still be regarded as "human body"? Does the memory information after the transplantation and enhancement belong to the "mind" category? What is the comparability or equivalence between the technical, artificial development and the natural development of human? Which is the comparability, for example, the comparability of effect and function, or attribute and motivation or goal? There is no doubt that it needs more research, particularly in light of the philosophical issues in the development of artificial intelligence.

Furthermore, the discussion above extends to a fundamental question of axiology, *that is*, if everything above is possible, do people want to be digitalized? In particular, if atoms and bits are "freely convertible", real and digital information can be arbitrarily converted, people as the subject can also be virtualized and turn their bodies into digital information and images, just like turning their bodies into souls.

When people become a kind of beings without illness, death and the limits of time and space through the mental cloning, is it still a "life"? What is the meaning and significance of life in this situation? Are we willing to make technology develop in this direction? Are people willing to completely eliminate their real "flesh bodies" that allow them to enjoy all kinds of "earthly" happiness with all kinds of pains and limitations? Is "removal of human body (carnal)" a valuable choice? Can this be the consensus of the vast majority of people? In other words, in the problem of relationship between IT and the significance of human life, we will undoubtedly face the value predicament, which is the same as the one that cloning technology faces. If we stop all the imagination of improving our well-being through IT because of the sanctity of human dignity, we may lose an important way to make people happier. If this kind of technology is unrestricted and free to carry out, it does have unpredictable risks, especially "humanistic risk". Therefore, the PIT, including the IT ethics, undoubtedly needs to intensify the exploration of such issues with sharp antagonism as well as infinite possibilities.

13.3 Some Frontier Issues in the Philosophy of Information Technology

Modern information technology is the fastest-growing technology. It has continuously expanded into new frontier fields with a fascinating future trend. This has also triggered new philosophical issues and promoted new philosophical thinking. Here, we select four frontier issues including the Internet of Things (IoT), artificial emotions (AE), artificial experience, and the interface of knowing and doing to explore the philosophical significance of its latest development. From these frontier issues, it is manifested that the information technology philosophy is an open system and a philosophical study that is constantly driven by the "frontier paradigm".

1. **New exploration of IoT and network philosophy**

As the iconic technology of the information age, the Internet has unlimited development possibilities since its birth. The IoT is currently being seen as a greater development of Internet functions and fields, surpassing the Internet in many ways. In other words, it makes the Internet extend in three dimensions, i.e. one is to extend from the interconnection of information to interconnection of things, the other is to extend from tracking things to controlling things, the third is to extend from the interconnection of artificial objects to natural ones, finally realizing the "Internet of Everything", *that is*, to achieve the interconnection of anything with anyone at anytime and anywhere. The important social and humanistic significance of the IoT can be seen from its generation of new ways of production, existence (things become "smart" objects with "things' position") and life (people are only a node in the ocean of the IoT). The IoT realizes the integration of the material world and the mental world (materials are endowed with "consciousness" and can "communicate" with human beings).

The emergence and use of the IoT have also raised new philosophical issues. Are people in the IoT getting new liberation or falling into new alienation? In the IoT, human is both subject and purposes, as well as objects and tools. In the process of controlling "things", people have achieved ubiquitous perception and monitoring through the IoT, and become subjects with higher degrees of freedom. But at the same time, people per se are also regarded as a type of "thing" which is placed in omnipresent perception and monitoring, and their freedom will be completely lost. When people and things are bound by the IoT, it leads to a Heidegger-style philosophical question, *that is*, Is there an additional "Ge-stell" for people and things? Both human beings and things will become "Bestand" thoroughly, "Heraus-fordem" (forced) by the IoT and turn into "resources" controllable by the monitors.

In the advanced stage of the IoT, when people can realize the ubiquitous perception, monitoring and "dispatch" of everything (including other people), the world we face may become the one that we can completely control, thus the world would lose all its "mystique" and be completely "disenchanted", not only in the epistemological sense, but also in the practical sense. The world is no longer characterized by "nature" in which "objects are gathered together by internet", and "objects are controlled by internet". When everything is included in the IoT as a technical form, when everything can be changed from the things in the natural network to the things in the technological network, the land becomes i-LAND, and the sky becomes i-SKY, everything completely lost their "freedom", and all become a kind of "factitious", "artificial", "electronic", "digital" or "technical" existence, the essence of "network" would be extended to the extreme. Internet resembles the black hole that "engulfs" everything and "sweeps" everything. To the extent, nothing could avoid the fate of being "Ge-stellized" by technology. Does this mean the complete loss of the naturalness of things, and people will have entered the era of the so-called "end of nature"?

Moreover, what would that be like if men possessed the capability of "interconnecting things" to incorporate everything into the IoT we built? It might be impossible for the scope of interconnecting and to what extent things can be interconnected to expand recklessly along with technological potentials. People will always set limits to the scope of interconnecting for political, humanistic and economic reasons. Of course, the specific setting of these limits would undoubtedly become an issue that is likely to cause debates and worth discussing. It can also be regarded as a question of "the philosophy of information technology in the future".

2. **Humanistic dimension of artificial emotion and information technology**

Artificial intelligence is an important field of contemporary information technology. In general, it mainly simulates the rational thinking activities of human beings that can be formalized and symbolized. However, artificial intelligence is now surpassing this range and expanding to the "irrational" field of "emotion", thus creating a new frontier of "Artificial Emotion" (AE).

AE originates from the study of emotional computers. One of the important concepts is "affective computing", which refers to developing a system with the ability of detecting, classifying, organizing and responding to human emotions. The computer is capable of recognizing and processing emotional information in the course of interacting with people, and can make intelligent, sensitive and friendly responses to human emotions. AE is a natural extension of "emotional computing". It refers to using artificial methods and techniques to imitate and extend human emotions, so that the machine can understand people's happiness and sadness with according acts. The "social robots" that have already existed are initially equipped with such functions. In connection with these functions, AE also studies machine algorithms that suppress bad emotions and explore ways to make the function of emotion in decision making by machine available. Mainly through simulating the control mode of human brain, it tries to build the mathematical model of feeling, perception and emotional decision behavior (human brain control), establish the machinery algorithm of emotional training, and even put forward machine realization strategy for inspiration (sudden enlightenment) (Wang, 2007).

Not only does AE have original meaning mentioned above, but also has "extended meaning", *that is*, the extension from endowing computers with human emotions to providing human beings with artificial emotions. These artificial emotions could be produced by supercomputers and "implanted" back to human brains. This is a new way to supplement, enrich and improve the contents and controlling ability of human emotions by feeding back artificial emotions produced outside to human beings per se. For example, extending this technology to human beings can be envisioned to help people's natural emotions, thereby enhancing people's ability of emotional expression, especially to help those who have flawed expressions fully express their emotions. Expression disorder or expression defect of emotions can be treated. In addition, there would be emotional artificial manufacturing, which means finding suitable technical means to construct a physiological or physicochemical state in the human brain that can produce the emotional state we need according to the emotional generation in the machine system. When this technology becomes convenient, it can make people's emotions that are difficult to generate in a natural state come to life with their own needs. The positive significance of artificial emotion technology lies in overcoming the individual emotional scarcity, solving the humanistic dilemma of the society, and opening up a new world of people's all-round development. Artificial emotional technology enables new emotional characteristics, types and expressions to be excavated constantly, further enriches people's emotional life and satisfies the growing emotional needs of human beings. Like the "knowledge innovation", "emotional innovation" becomes the spiritual condition that is necessary for human survival and development, providing new services for the new emotions that human's desire. From a philosophical point of view, AE is a kind of demand to realize the humanistic realm in the scientific and technological way, and thus becomes an indispensable component of human freedom and comprehensive development.

Of course, AE will also make us expose to new philosophical problems. The first one is concerned about the authenticity of emotions, specifically, is AE real emotion? If not, when AE can be created at will, "pseudo-emotions" will engulf all places of communication inevitably, and the true feelings of people may be more difficult to grasp. Further, it may cause the proliferation of unnatural, disguised or false emotions. Not only do people lose their inborn nature because of physical technology, but also at the core level of the spirit, i.e., at the level of emotion, lose the inborn nature, so not only does technology manipulate our material existence, but also dominates our spiritual and emotional survival. The second is the problem of individuality and richness of emotions. If artificial emotions originating from the same machine are introduced to different people, will they submerge the difference with the sameness, leading to emotional convergence of people, or homogenization? Is it necessary for emotions to converge? If there is a need for convergence, what the "standard" emotional pattern is? Emotional convergence may also be manifested in the process of technically eliminating negative emotions. Once a negative emotional or emotional state is eliminated "artificially" or "technically", positive emotions will not exist. The value of positive emotions and the meaning of its existence can only be found in the comparative experience of positive and negative emotions. This is also the limit of artificial emotions. It means that when we serve ourselves with artificial emotions, we must also be alert to the excessive dependence on artificial emotions in emotion recognition, expression and manufacturing, and avoid the loss of original emotion capacities, as over-dependence on tools in memory and calculation will result in degradation of memory and computing power. Therefore, the tension between human nature and technology is also sharply expressed in the development of artificial emotion technology.

3. **The technical generation of experience and the expansion of information technology in the spiritual world**

The way information technology involved in the process of cognition is diverse. Chip implantation and neural manipulation are the most attractive ways. They have the potential to "technically synthesize" part of the experience, so that our experience can be obtained from another type of source. We will have a new type of experience which is "artificial experience". The technical synthesis of experience provides another type of source of cognition and new insights into understanding the "source of cognition".

Experience often comes from the real experience of people, such as, behaviors, practices, activities, etc. in the real world. "Hands-on experience" is "personal experience" with a similar meaning to the "life history". We can consider the experience that is not one's physical or practical experience as a "second-hand experience" or "indirect experience", such as learning the local customs of a certain place through books or words, and a kind of "indirect experience" is a non-experienced experience. "First-hand experience" is distinguished from "second-hand experience" in that the former is the experience of direct contact with the object or event when the body is present, and the latter is gained experience with the object or event when the body is not present.

Experience as a mental image is undoubtedly related to the specific state of the human brain. There are many ways in which this particular state can occur. Different types of experiences may generate different types of experience, and further form different sources of experience. With modern information technology, we are increasingly expanding the source of experience, *that is*, using technology—especially contemporary information technology—to help us generate experiences, including the stimulation of virtual environments, the transplantation of memory materials, and the construction of neural networks. They form an evolutionary chain from primary one to advanced one that are embedded in our cognitive process and evolve into another important source of experience.

The mind and cognition are essentially embodied as the appearance of the overall activity of the nervous system. As psychic phenomenon, experience is undoubtedly associated with specific state of neural connections. Nowadays, by virtue of correlating the location of neural activity to certain conscious experience, cognitive neuroscience is trying to find the neural correlates of consciousness (abbreviated as "ncc"), and then integrate all nccs to form uppercase NCC, we can obtain a whole map of nerve location to consciousness. By interpreting and simulating the working mechanisms of neurons, these results can be used to manually manipulate the nervous system of human brain to "weave" the required structure and generate corresponding mental content, including specific experiences. Thus, figuring out the neural correlation of experience, constructing this correlation technically as well as recreating the neural correlation of one person's experience to another one becomes an important way of technically generating experience.

For the technical generation of experience, some theories in epistemology and psychology provide a philosophical basis for explaining its possibilities, while cognitive science and technology are exploring ways to achieve the technical generation of experience. From the perspective of materialism, experience adds a new way of generating, and the artificial approach goes beyond the natural approach, so that acquired experience can be generated by technology in addition to natural formation. The source of experience can be divided into "conventional sources" (undergoing, learning, reflection, behavior, practice…) and "unconventional sources" (implantation of memory materials or artificial construction of related nerves in the brain). This also forms a new classification of experience, i.e. experience of priori theory, experience of theory of reflection, experience of technology or manufacturing theory. If the technical generation experience is realized through the artificial construction of the neural network, then there will be more the "tool" or "intermediary" component in the reflection. In other words, the "brain-machine" interaction in the reflection will be much stronger. Whether it is the "inter-brain" or "brain-machine" type of experience enhancement, it can be regarded as the enhancement of the ability of the subject of the experience, and finally attributes to the improvement of people's cognitive ability. This also shows that if the past experience is the unity of the state of mind and the state of practice alone, then it can also be the unity of the state of mind and the state of technology, and thus further indicates that the state of technology and the state of practice can be blended substantively.

Generating experience through technology can greatly expand the world of human experience by overcoming limitations of gaining experience through the conventional, direct, personal experience, and allow the world in which we live to extend beyond our direct experience, even go beyond the world of possible experience. We also perceive the space field we can never reach in person, thus enriching people's experience, or making more people become experienced people, thereby helping inexperienced people form more "real" life experiences with corresponding effects. This also means that the technical generation of experience can change the traditional way of learning and cognitive process. For example, the skill experience can not only be learned through a long and arduous imitation process, but also be quickly supplemented and improved in a short time. In addition, when the manual operation of experience can be carried out according to people's needs, when "remembering" and "forgetting" are equipped with high technical operability, it is easy for us to recall the experiences we have forgotten, to consolidate precious experiences, to get rid of painful memories and experiences, and to heal the psychological and soul traumas. As a result, the quality of our mental life can be improved.

In short, the technical generation of experience is another source of experience. It changes our ways of experiencing and enriches the world of human experience. The technical generation experience has given the experience a new philosophical meaning, so that the traditional "direct experience" and "indirect experience" can be integrated in a new type of experience, and even the spiritual philosophy has new problems and new enlightenment. This can promote the intersection of philosophy of information technology and epistemology, cognitive science, and mental philosophy in the future.

4. **Brain-computer interface and the unity of knowing and doing**

Human-machine interface is the interaction between human and machine. The change of human-machine interface is the change of human-machine interaction mode. In the human-machine interface, from the computer keyboard to the brain-computer interface (BCI; or brain-machine interface), the trend is becoming more and more complicated with the special function of "knowing-and-doing interconnection" derived.

The BCI can be regarded as a means for the computer to understand the human brain. It is another important dimension in the development of human-machine interface technology. At present, the human-machine interface is mainly a limb-operated interface, a "hand-machine interface". Developing BCI technology opens up another model of human-machine interaction, the information exchange between the computer (interface device) and human. This can be realized by importing human brain waves into the machines.

The BCI is a direct connection path established between the human brain or the animal brain and external devices, where the physical representations of the activities in the brain (such as brain waves) are obtained by a computer or a corresponding technical device so that to some extent the content of these brain activities can be understood and applied. When the BCI develops to a certain stage, people can use their own "thinking" (brain waves) to control the computer and other devices connected to it, forming effects similar to "action" or "behavior". Of course, when constructing a BCI with guiding actions, how to separate the idea of action from brain waves is the most difficult and most critical part. Various signal processing techniques are needed to detect the action idea, and then the measured original action idea is converted into instructions that the machine can understand. To achieve this function with BCI technology is not out of reach, because some laboratories have used it to collect EEG signals from physically disabled people and then use these signals to operate their prosthetics. There are also labs that use human brains to manipulate robots to play football.

Depending on a BCI device, a person can make a physical change of an object without any physical action by his hand. This is the extension of the function from the BCI to the interface of knowing and doing. Its main technological process is reading the machine information reflected the state of human thinking, mind and ideas, forming operational instructions of automatic machine, and undertaking the physical movement by actuation system. Its aggregate form becomes the action of substituting human in function. This is called "alternative human action" or "human practice in the sense of extension".

Once the BCI transmits the behavioral intention and instruction from the brain effectively to the information execution system, it will cause the actual operation and change of the object, as it has the effect of "doing something". The technical system between the human brain and the object can be referred to as the "knowing and doing interface system". It is a human-machine interface that does not have to rely on the action of human body (except the brain). A human-machine interface is no longer a form of information transformation alone, but also a tool for interconnecting the informational world and the material world. In a sense, one's thinking of doing something can directly transit into the "practice" or "action" stage, forming an interconnection between "knowing" and "acting" (doing). However, the practices and actions have been informatized deeply to become "informatizational practice" and "informatizational action".

From the perspective of relativity, knowing-and-doing interface is the extension of human-machine interface from the information field to the behavior field. It can also be regarded as the technological connection between mind and the material world. This interface device can produce artificial action outside the human body, as long as it can read the mind or the intention of the human brain. At this point, the external devices driven by the knowing-and-doing interface resemble "artificial limbs with nerve". The continuous progress in the direction of knowing-and-doing interface indicates that the special human action can surpass the human body. In addition, the integration of human brain activity and anthropomorphic action of the functional device indicates that the "thinking" and "acting" can be integrated into the human-machine system. So that the traditional "doing by hand" and "doing by word" (instructing others to do things or "telling the machine to do things") can be substituted to "doing only by thinking". Using BCI to control the intelligent machine to do what we want is the so-called "only thinking, things can be done".

The BCI is the advanced form of human-machine interface, and the knowing-and-doing interface is a functional description of the BCI. As for human activities, "knowing" is a kind of informational movement, and "doing" is a kind of physical movement. The knowing-and-doing interface connects the two movements and carries out the latter movement under the control of the previous movement. It externalizes thinking activities in mind without the body's physical movement. It is the "shift" of the knowing-and-doing interface. This shift also allows us to turn to the outside world so as to further understand how knowing and doing are connected and transformed. Although there is debate about whether the physical movement of the technology system driven by knowing-and-doing interface belongs to human action, it is obvious that only the idea per se cannot achieve the effect of doing something, so at least it can be considered as an extended human action, or so-called "extended practice", just as there is extended cognition. In such a limited understanding, people, if there is knowing, will transmit their intentions to the device of doing, so that the device will be controlled under "digitalized" instructions (a kind of knowing), and then creates artifacts or produces corresponding movements. From that, it can also be seen that people are constantly creating new interfaces, distinguishing different fields, and then considering how to connect them through interfaces.

In short, the frontier of modern information technology should be viewed as a new philosophical object, and new philosophical ideas can be generated continuously, so that the philosophy of information technology and philosophy of technology to obtain a new growth point.

References

Brey, P., & Soraker, J. H. (2009). Philosophy of computing and information technology. In D. Gabbay, P. Thagard, & J. Woods (Eds.), *Handbook of the philosophy of science* (pp. 1341–1408). Amsterdam: Elsevier.

Mitcham, C. (2004). Philosophy of information technology. In L. Floridi, (Ed.), *The Blackwell guide to the philosophy of computing and information*. Blackwell Philosophy Guides (pp. 327–336). Blackwell Publishing Ltd.

Wang, Z. (2007). *Machine intelligence: Artificial psychology* (pp. 9–10). Machinery Industry Press.

Xiao, F. (2012). On the philosophical meaning of information technology. *Journal of Northeast University (Social Science), 14*(4), 283–288.

Some Further References in Chinese

Xiao, F. (2004). The digital development of human being and the new subject of hominology. *Jiangxi Social Sciences, 2,* 13–19.

Xiao, F. (2008). Towards the philosophy of information technology. *Studies in Dialectics of Nature, 1,* 40–45.

Xiao, F. (2012). Artificial emotion: The new integration of technology and humanity. *Marxism & Reality, 1,* 91–97.

Xiao, F. (2013a). Information, information technology and informational epistemology. *Journal of Changsha University of Science & Technology(Social Science Edition), 1,* 5–10.

Xiao, F. (2013b). From internet to internet of things: A new exploration of philosophy of technology. *Journal of Northeastern University (Social Science), 3,* 221–227.

Xiao, F. (2014a). The interface of knowing-practicing and its philosophical meanings. *Journal of Northeastern University (Social Science Edition), 2,* 111–118.

Xiao, F. (2014b). Technical generation of experience. *Marxism & Reality, 4,* 54–60.

Xiao, F. (2014c). The rise of the philosophy of information technology. *Studies in Philosophy of Science and Technology, 6,* 45–52.

Xiao, F. (2015). Philosophy of information technology: New trend of philosophy of technology. *Journal of Changsha University of Science & Technology (Social Science Edition), 3,* 5–11.

Xiao, F. (2016a). *Philosophy of information technology*. South China University of Technology Press.

Xiao, F. (2016b). Information technology in the view of "Being". *Studies in Philosophy of Science and Technology, 1,* 65–72.

Xiao, F. (2016c). Internet of mind and its philosophical analysis. *Journal of Northeastern University (Social Science Edition), 4,* 331–337.

Xiao, F. (2017). *Information technology in the view of philosophy*. Science Press.

Yi, X., & Li, X. (2017). The research on the epistemology of information technology philosophy: Current situation and commentary. *Science and Technology Management Research, 1,* 257–261.

Xu, Y. (2016). From "Information philosophy" to "Information Technology Philosophy"—A research route turn with nominalism. *Journal of Wuhan University of Science and Technology (Social Science Edition), 12,* 589–602.

Zhou, L. (2004). Four basic questions in the study of philosophy of information technology. In Z. Liu, et al. (Eds.), *Research yearbook on philosophy of technology in China in 2003* (pp. 165–174). Dalian University of Technology Press.

Chapter 14
Approaches of Nano-ethics and the Chinese Perspective of Ethics of Technology

Guoyu Wang and Lei Li

Abstract In this article, three main approaches of nano-ethics were analyzed from the view of ethical theories and methods, in order to explore the key issues in ethics of high-tech. The first approach is the reduction of ethics as a "heuristic of fear" that serves as a guide to reflect on what the human truly valued is, to appeal for suspending nanotechnology development; The second one is the reduction of ethics as consequentialist evaluation and calculation, which takes the ethical issues as accounting of the risks and benefits and whether man are willing to accept these risks. And the third one is putting nanotechnology in a certain social and cultural context, trying to open the "black box" of the development of nanotechnology, which reveals the close relationship between technology and society, in order to better understand the relationship between nanotechnology and social cultural values. However, these approaches are still difficult to solve the problem of co-existence with uncertainty. Based on the analysis of the approaches above, this article puts forward the conditions of future-oriented nano-ethics framework from the Chinese perspective, utilizing from the ideological resources of the coexistence with uncertainty in Chinese philosophy.

Keywords Approaches of nano-ethics · Uncertainty · Chinese perspective · Strategies · Introduction

Since the 21st century, with the rapid development of nanotechnology, biotechnology, information technology and cognitive science (the so-called converging technologies) and so on, the ethical, legal and social issues that may be generated

---

G. Wang (✉)
School of Philosophy, Fudan University, No. 220, Handan Road, Yangpu District, Shanghai, China
e-mail: wguoyu@fudan.edu.cn

L. Li
Department of Philosophy, Dalian University of Technology, No. 2, Linggong Road, Ganjingzi District, Dalian, China
e-mail: lilei.1988@hotmail.com

Q. Wang (ed.), *Chinese Philosophy of Technology*, Philosophy of Engineering and Technology 34, https://doi.org/10.1007/978-981-15-1952-9_14

by the development of high-tech will become increasingly urgent and prominent. The basic theories and methods of ethics of technology are difficult to deal with the increasing risk of high-tech and uncertain consequences of application in an effective way. Taking nano-ethics as an example, it has been more than 10 years since the term was proposed. In this process, the debates about nano-ethics with the focus on the safety of nano-materials, nano-devices and personal privacy, human enhancement in bio-nanotechnology, the fair distribution of benefits and risks of nanotechnology are fierce. From the perspective of ethical theories and methods, there are three main approaches to nano-ethics. Although these approaches are helpful to analyze the risk and uncertain consequences of nanotechnology, it is difficult to avoid the defects in dealing with the multiple value conflicts and maneuverability in technical activity, and find a feasible action strategy. However, in the context of Chinese philosophy, combined with the practice of China's high-tech research and development, we propose an idea of ethics of technology based on harmony and strive to build a dynamic and feasible action framework. This will help to solve the current theoretical and methodological problems facing the ethics of technology.

14.1 Three Approaches of Nano-ethics

1. **The Heuristic of Fear**

The ethical reflection on nanotechnology originated from people's fear and worry about nanotechnology. In 1986, Eric Drexler published the book *Engines of Creation*, in which he put forward the core concept of self-replicating machines. His self-organized nano-machine inspired by John von Neumann's self-replicating machines and Feynman's early view on the possibility of molecular manipulation. A molecular assembler is a molecular machine by which we can use a program to build any molecular structure or device from the simple chemical component. This nano-scale assembler can place molecules in any way, construct chemically stable structures, and be able to create a copy of their own set by the program. Starting the self-replication will be likely lead to an exponential growth of existing assembler, and can construct a macroscopic object. Drexler was clear about the result, "we will be able to rebuild or destroy our world by using the assembler" (Drexler, 1986), and the "engines of creation" may also be the "engines of human destruction". The concept of Drexler's nanotechnology also led to utopian dreams and dystopian nightmares and dominated the later debates on ethics (Gordijn, 2005). In 2000, American computer engineer Bill Joy published "why the future doesn't need us" in the *Wired Magazine*, which especially pointed out the destructive force brought by the combination of nanotechnology, computer technology and gene technology. "The devastating self-replicating power in genetic engineering, nanotechnology and robots is likely to stop the development of our humanity abruptly" (Joy, 2000), and the risk is far more than weapons of mass destruction including nuclear weapons, biological weapons, and chemical weapons. In 2002, the American best-selling author Michael Clayton published the book *Prey* (Crichton, 2002), which brought the self-replicating nanotechnology into the public view again, showing the devastating effects on the biosphere caused by a group of nano-robots escaping from the lab. Since then, the fear and resistance to nanotechnology has spread all over the world. International environmental organization ETC (Action Group on Erosion, Technology and Concentration) states for "nanotechnology will lead us to the abyss" and "No Small Matter" and a moratorium on environmental emissions and commercial applications of nano-materials, and the prohibition on self-assembling nanomaterial and patents issued for nanotechnology (ETC Group, 2013).

It can be said that from the fear of nanotechnology to the appeal of suspending nanotechnology research and development, this per se is an action that contains people's value assessment. The reason why people refuse nanotechnology is the fear of the possible dangerous consequences. Because of the uncertainty of nanotechnology self and consequences of nanotechnology application, people's evaluation on it, to a great extent, is based on their intuition and emotion. This is what Hans Jonas calls "the heuristics of fear". In Jonas's view, it is important to have a non-logical and non-methodological perception to avoid something when we do not know the danger, and we do not know what should be kept and why to do so. It is more direct and urgent to make us know the "evil" that we do not want and confirm the "good" that we want. Therefore, when referring to moral philosophy, we should learn from our fears first in order to know what we really cherish (Jonas, 1984). And fear is a better guide.

2. **The Consequentialist Evaluation**

To a large extent, the debate on the benefits and risks of nanotechnology is often reduced to the presumption of its possible consequences even with too much futuristic color (Nordmann & Rip, 2009). The debates on ethical issues of nanotechnology imperceptibly fall into the pattern of consequentialist arguments, while the consequentialist arguments, in large measure are difficult to avoid in that the technology development always takes a special way of persuasion, *that is*, if we invest in new science and technology, it will enhance our knowledge and ability for the manipulation of nature, and when the knowledge and ability have a positive effect on humans, they will ultimately enhance the overall happiness.

However, when the consequentialist argument is applied to nanotechnology, there arise three challenges. Specifically, first, the prospect of the future of nanotechnology at this stage lacks factual confirmation, for example, nanotechnology will help interface the human brain and the computer, so we can learn French only by implanting a chip. Second, the benefit-cost ratio of nanotechnology is not confirmed. It is not only a scientific problem, but also involves the perception of risk of different stakeholders, such as supporters and opponents of nanotechnology, the experts and the general public. Third, whether the future of nanotechnology is really beautiful as promised is related to different value judgments on different stakeholders. The three challenges also make the consequentialists' evaluation in trouble since the typical consequentialism has strong empiricist orientation and takes the result of current behavior as the only evaluation basis. However, the uncertain consequences of nanotechnology make this evaluation lack factual confirmation.

At present, either from the perspective of deontology or utilitarianism, the research into ethical issues of nanotechnology is always dominated by consequentialists. These ethical issues arose from the possibility of appealing nanotechnology are often equal to possible ethical consequences. In the early articles on nano-ethics, most authors provided a series of possible conflicts of interests and extreme judgment by either appealing to the analysis of consequences (consequentialist) or putting forward values generally accepted (deontological) (Ferrari, 2010). For example, in the respect of bio-nanotechnology and nano-medicine, the developments of nanomaterial make better biologically compatible medical implants (such as orthopedic implants, heart valves, etc.) possible, and the developments of bionano devices and biosensors make organ transplantation and replace (such as artificial retina, artificial cochlea) and more efficient information processing and storage possible. These possibilities generate not only the improvement of human interests and quality of life, but also a series of ethical and social issues, especially human enhancement (Jotterand, 2008). James Moor and John Weckert pointed out that a person's brain operation theoretically can be tapped through nanotechnology and wireless transmission without his knowing. It may be difficult to read his thoughts, but quite simple to catch the specific indicative mental state (Moor & Weckert, 2004). In other words, nano-devices or nano-scale implants may cause violations of personal privacy or the control of others' actions and ideas. With regard to the social influence on human enhancement with nanotechnology, Patrick Lin and Fritz Allhoff point out that we can expect the enhanced person is more productive and creative, but it is difficult to calculate whether these benefits will outweigh its costs or risks (Lin & Allhoff, 2008). From the examples above, like all the consequentialists' arguments, human enhancement with nanotechnology has two major defects. Specifically, on the one hand, it inevitably involves utilitarian trade-offs between risks and interests, and it is difficult to reach a consensus in a multi-valued world because ethical belief is different from negotiable interests. On the other hand, in view of the uncertainty of nanotechnology, ethical evaluation based on supposed consequences inevitably goes to utopian or dystopian fantasy which is harmful to the development of nanotechnology.

Due to the uncertainty of nanotechnology, the deontological and utilitarian analysis of ethical issues of nanotechnology has no unanimous conclusion can be drawn. Thus, a new trend emerged in nano-ethics, namely from the perspective of technology assessment, the uncertain consequence is themed as the risk of nanotechnology. What is the risk of nanotechnology? Should we accept the risk of nanotechnology? Furthermore, the risk is regarded as the only problem emerging from the application of nanotechnology, and other problems have been redefined as the magnitude of risk. In other words, the issues of, fair distribution, enhancement and the needs of the public to be fully informed are re-described as part of the traditional risks and side effects of technological development. The consequence is that the main ethical issues are only regarded as the willingness (or unwilling) to accept these risks (Ferrari, 2010). The risk of nanotechnology is widely discussed in literature of technology policy and risk management, which is the manifestation of this tendency. For example, Dan. M. Kahan point out that it is necessary to systematically study the public perception and attitudes towards the risks and benefits of nanotechnology in order to promote the public consultation on nanotechnology (Kahan, 2009). Therefore, the ethical issues of nanotechnology have been transformed into the risk assessment of nanotechnology to a great extent.

3. **The Contextualist Prudence**

The assessment of nanotechnology confronts a new epistemological situation characterized by uncertainty, ignorance and fuzziness. These features cannot be fully grasped by assessment. With regard to the ethical reflection on nanotechnology, we must go beyond the consideration of its consequences, put it in a specific context and pay close attention to the conditions, backgrounds, processes and ways of its development. Only in this way can we grasp the rich meaning of the challenges posed by nanotechnology. "Moreover, the nanotechnology-related ethical issues identified to date have not been engendered solely or predominantly by intrinsic features of the nanotechnology field. Rather, their genesis is also due to contingent extrinsic features of nanotechnology's microscopic, mesoscopic and macroscopic social contexts" (McGinn, 2010). Joachim Schummer also indicates that the novelty of nanotechnology lies not in its technological particularity, but in its perspective of the role of science and technology in the society. The uniqueness of nanotechnology is that it takes a unique form which is playing with the socially built boundary (such as life and non-life, natural and artificial, and even natural science and social science) (Schummer, 2009). Therefore, from the perspective of STS, we can get rid of the plight of the consequentialist assessment of nanotechnology risk by understanding the social and cultural context generating nanotechnology.

In view of the uncertainty of nanotechnology and cognitive limitations, ethics can play a role as sophisticated forms of prudence. This is reflected in focusing on the cultural significance and social embedded of nanotechnology innovation, the motivation and scope of nanotechnology development and thinking how to develop nanotechnology sustainably and responsibly. It can be said that the ethical issues of nanotechnology are no longer understood as trade-offs between risks and interests, but rather an attempt to deal with the uncertainty of nanotechnology in an ethical way (Ferrari, 2010). As is stated in a report of UNESCO that we should reflect on the structure of science, public trust and responsibility (UNSCO, 2006), the European Commission proposes that nanotechnology development should follow the principles of meaning, sustainability, inclusiveness, excellence, innovation and accountability in its code of conduct for responsible nano-sciences and nanotechnologies research (European Commission, 2013). These prudent action guides are concerned with the complex and integral dimensions of the challenges raised by nanotechnology. In addition to calling for responsible development, some scholars reflect on the uncertainty of nanotechnology from the epistemological and social perspective, and point out that we should deal with risk assessment from a broader perspective. For example, R. Schomberg stressed the importance of "collective responsibility", especially individual responsibility and supra-personal assessment mechanism in unexpected consequences and collective decision-making (Schomberg, 2007).

Other scholars put forward that we must develop new forms of ethical reflection, and focus on the historical and philosophical sources of nanotechnology, and the position of science and technology in society. Only in this way can we grasp the complexity of the problems raised by nanotechnology. Jean-Pierre Dupuy and Alexei Grinbaum also believe that the real challenge of nanotechnology is the uncertain dimension of its risk. In order to fundamentally understand the uncertainty of nanotechnology, we cannot stop to investigate its possible consequences, but explore the metaphysical research program to motivate the research and development of nanotechnology. This method is committed to clarify the vision, idea and view of nature represented by nanotechnology through exploring the social and cultural dimensions of the image hosted by nanotechnology. Only in this way can we get a different pattern of ethical reflection and a kind of ethics beyond prudence, namely the "ongoing normative assessment" (Dupuy & Grinbaum, 2004).

14.2 The Theoretical and Methodological Problems of Approaches of Nano-ethics

From the analysis above, it can be seen that the three main approaches of nanoethics are based on different ethical theories and methods. There are clear conflicts among the three approaches, and it is difficult to reconcile the differences between each other.

It is notable that the enlightenment of fear has obvious emotionalist and intuitionist characteristics. Although the moral judgment appealing to people's emotion and intuition can arouse their attention and sympathy in a certain extent, it also has to face criticism featured with subjectivism and non-cognitivism too much. The consensus of moral judgment in the enlightenment of fear is guaranteed not by certain rational approach, but by non-rational influence from emotion, attitude, etc. In this way, the moral judgment of specific technology is likely to become an expression of personal subjective and personal likes and dislikes, which lacks the analysis of the background and the temporal and spatial characteristics of fear, and then, this may further disintegrate the objectivity of moral judgment. Although from a personal perspective, it is also a self-preservation method to know some consequences that we do not expect and need to avoid by means of fear when we cannot determine the possible consequences of nanotechnology; from the viewpoint of the whole society, the development of nanotechnology closely related to the political, economic and cultural systems is not determined by a single person but a social action to a large extent. How to avoid the possible catastrophic consequences? Turning to fear is of no avail. The subjective color of the enlightenment of fear often limits the discussion to utopian or dystopian fantasies.

The main difficulty of the consequentialist risk assessment is that it involves forecasting the development of science and technology. This forecast not only faces the cognitive deficiency in the inference of the future, but also faces the uncertainty of its possible consequences. As Nathan Rosenberg indicated, new technologies come into the world in a very primitive condition and with properties and characteristics whose utilities cannot be immediately appreciated, potential uses, complementary innovations, systems integration, problem-solving myopia, passing the needs test, competing with the past, etc., make the forecast of new technology uncertain (Rosenberg, 1995). Therefore, the risk assessment of nanotechnology, including all kinds of existing risk assessment methods, is faced with great difficulties. The consequentialist risk assessment framework appeals to the condition that the presumed potential negative effects have been identified, but it is impossible to quantify these risks in the discussion in that there is a lack of relevant data or an understanding of its uncertain and inaccurate nature. Even if we are able to quantify these risks, the significance of ethics is also problematic. As Jean-Pierre Dupuy says, "if ethics were a calculus, even if this calculus turned out to be impossible to undertake in practice, every ethical problem could be presumed to have a solution, whether or not we were capable of determining it" (Dupuy, 2007). Therefore, it is difficult to get rid of the consequentialist influence on the ethical evaluation and risk assessment of nanotechnology. The consequence uncertainty of nanotechnology makes either ethical assessment or risk assessment difficult to work. The consequentialist assessment is not only faced with the cognitive deficiency, but also difficult to have a positive impact on the development of nanotechnology.

The prudence based on contextualism attempts to open the "black box" of the development of nanotechnology and to show the close relationship between nanotechnology and society. The studies on the contemporary philosophy of technology have shown that technology development is not linear and completely independent, but a kind of social construction under the framework of technology and social systems. In other words, the development of technology and society is co-evolutionary, to a certain extent, is a random outcome of social interaction. The development and design of specific technological products appear in the social field, involving different actors' different interests, goals, resources and powers. With the help of economic, sociology and other social scientific methods, we can clarify the relationship between technology development and related social factors and better describe the conflicts between technology and social cultural factors so as to provide the basis of action for making technology policy and managing and communicating the conflicts and risks of technology. There are many similar studies in this area, such as Schummer's, Bincent's and Gaskell's. Schummer studied the social cultural factors like language, cultural inheritance, economy and politics in order to understand how the perception of the ethical issues of nanotechnology was influenced (Schummer, 2006). Bincent distinguished the two kinds of culture of nanotechnology (Bensaude-Bincent, 2004).

Gaskell et al. studied the cultural supportive factors behind the nanotechnology innovation in European and American public sphere (Gaskell, Eyck, Jackson, & Veltri, 2005). However, technological conflicts are not only the conflicts between technological means and tools, but also about the conflicts of future vision, human nature features and social planning, so the technological conflicts have moral implications (Grunwald, 1996). If we only make descriptive efforts, we are unable to grasp the normative dimension of the technological conflict. If the normative dimension leads to conflicts of moral beliefs of individuals or social groups, the ethical reflection on technology is essential. The prudence based on contextualism mainly reveals the relationship between the development of nanotechnology and the society from the perspective of STS and is expected to point out the way for the development of nanotechnology. However, the prudence does not grasp the moral connotation of technological conflict, and it often stays in static descriptive efforts without necessary normative reflection. Furthermore, the ethical reflection does not play its due role in dealing with the conflicts of nanotechnology development.

Nanotechnology is an uncertain technology. First of all, the uncertainty of nanotechnology per se includes its definition, scope and so on. As we all know, nanometer is a concept of scale, but the technological definition of this scale unit dissolves the boundaries between disciplines. Almost all disciplines have traces of nanometers. Scientists can study nanotechnology from different disciplines and perspectives, which directly leads to the uncertainty of research methods, so that people are also difficult to fully understand its characteristics. The second is the uncertainty of applications and consequences of nanotechnology. Nanotechnology, also known as "enabling technology", may improve existing technology in almost all areas and can be applied to almost any area from aerospace to daily life. Compared with the explicit and direct purpose of traditional technology, the purpose of nanotechnology is not clear. Especially in the so-called "converging technology", its purpose, direction and results of development are even more unpredictable. The third is the uncertainty of ethical, legal, social consequences of nanotechnology. To a large extent, nanotechnology is still a future and possible technology. All of the possibilities of nanotechnology may become reality, but we are not sure what kind of ethical issues will be raised because of cognitive limitations.

Due to the uncertainty of nanotechnology itself, its application and consequences, the enlightenment of fear and the risk assessment often fall into the speculation of the future possibility of nanotechnology, ignoring the specific technological development; and the interpretation of contextualism often stay in the static empirical description of the relationship between technology and society, lacking normative dimensions and the necessary contemplation of possibilities. Many scholars such as Alfred Nordmann (Nordmann, 2007) also criticized these problems. Based on this background, we believe that nano-ethics must break through the existing research mode, transforming itself from the critical reflection of emerging technologies based on traditional ethical norms to the creative construction of future-oriented, open, dynamic and normative nano-ethics.

14.3 The Chinese Perspective of Nano-ethics

The first condition of future-oriented nano-ethics is to coexist with uncertainty. It is at this point that the practical wisdom of ancient China can be an important source of thought for us to construct a new type of ethics. Different from the Western philosophy and ethics which pursue the certain conceptual system and universal principle behind the uncertain world, the Chinese philosophy does not presuppose an invariable, abstract noumenon and transcendental world, but directly faces the life and growth in nature. In the view of ancient Chinese philosopher, the whole world is a dynamic and open process, and they believe that change is the reality of the world. The task of philosophy, therefore, is to explore the practical wisdom of living with this ever-changing world. Although the Chinese philosophy also has metaphysical ontological inquiries, this ontology is still characterized by change. Everything is developing in the process of changing and transforming, and even "the *Dao* changes again and again". Therefore, in this sense, we can also say that the Chinese philosophy is about coexistence with uncertainty.

So how do we live with uncertainty? Ancient Chinese sages developed a set of feasible strategies for action, which not only took into account the metaphysical basis and moral principles of action, but also paid more attention to the strategic principles of weighing whether it is feasible or not, how to be feasible. In summary, the feasible strategies for action include the following aspects: Firstly, it is essential to respect and comply with the natural law. The "law" here can be understood as the *Dao* in Chinese philosophy. Of course, according to Confucianism, in addition to the *Dao* of heaven, there is the *Dao* of human, namely, social rules and ethical norms. In Chinese philosophy, *Dao* is the guiding rule of action, and is the essential condition of feasibility. Secondly, it is necessary to value the situation and seize the opportunity. The true way of respecting the *Dao* and acting according to it is to grasp the opportunity by considering the situation: "Keep still when it is time to keep still. Remain active when it is time to remain active. When action and resting do not miss their time, their way becomes promising and brilliant" (The Complete I Ching, 1998). Thirdly, "virtue" is a guarantee of the action. Human actions should follow not only the *Dao* of heaven, but also follow *Dao* of human. It is related to whether the behavior of the actor can be recognized and whether it is feasible.

The reflection on the feasible strategies for action in Chinese philosophy is of great significance to cope with the possible challenges in the development of nanotechnology today. We believe that if we want to get out of the intellectual and moral dilemmas facing to the study of the uncertainty of nanotechnology and its risks, we must seek and explore the boundary conditions of feasibility, that is, what the boundary conditions for the realization of possibility are. Considering that nanotechnology itself is a dynamic process, nano-ethics must consider the following conditions if it is to be able to "walk with science".

(1) To be open and dynamic. It requires us to adopt a future-oriented and open attitude, keep pace with the times, ensure the controllability and revisability of behaviors, and timely adjust development strategies on the basis of the development status of nanotechnology as well as the acceptability of nanotechnology in society.
(2) To be specific. This requires not to talk about the abstract nanotechnology at the level of abstract principles to discuss what to do and what not to do, but to go deep into the frontier field of nanotechnology, to analyze case by case, to determine what kind of nanotechnology should be developed and how to develop and govern on the basis of the different areas, different nanotechnology, different properties of nanometer materials.
(3) To be real-time. Not only does that refer to the time on the temporal level, but also includes the requirement of timing. Time is different from timing. The latter is a form, and the timing is associated with specific situation, *that is*, it is necessary to measure the situation and make real-time assessment and judgment according to the current situation of nanotechnology development.
(4) To be holistic. The Chinese philosophy regards heaven, earth and man as an interrelated integrity, and human behavior is closely related to the changes of the universe and the fate of the human being. From this point of view, the development of nanotechnology is not only a specific technical issue, but also a possible issue related to human interests and destiny. Thus, nanoethics should also consider both the interests of human beings and the influence of other species as well as the integral fate of human beings.

To cope with the possible ethical issues caused by the uncertainty of nanotechnology and its risk, ethicists and scientists must analyze the possible conditions, cultural environment of nanotechnology development, pay attention to the relations and interactions between various elements in the process of nanotechnology development, specifically, to identify ethical issues, assess possible consequences, social influences and social acceptability of nanotechnology and propose a feasible ethical framework for action on this basis.

14.4 Conclusion

At present, the research on nano-ethics is faced with two dilemmas: on the one hand, the "heuristic fear" from the mode of "if and then" is far from the actual development of nanotechnology, and it has obvious characteristics of emotionalism and intuitionism. Thus it is difficult to obtain the recognition and response from the scientific community. On the other hand, there is a gap between the development of nanotechnology and the research on nano-ethics. This makes the ethical research into nanotechnology not understand the practice and context of nanotechnology, and lack of necessary theoretical framework and methodology support. Therefore, the research of nano-ethics must be carried out synchronously between ethics and science. It is necessary to track and participate in the practice of nanotechnology and understand the scientific context and possible direction of the development of nanotechnology. Besides, we face the uncertainty of nanotechnology and cooperate with scientists to explore the ethical framework of the feasibility of developing nanotechnology.

Starting from the ideas on the practical wisdom in Chinese philosophy, along with the practice of nanotechnology development in China, we propose a Chinese approach of nano-ethics. It is believed that nano-ethics should proceed from the open, dynamic, specific, real-time and holistic strategic principles, and work with nanotechnology to promote and ensure the responsible development of nanotechnology.

References

Alfred, H. (1998). *The complete I Ching: the definitive translation from the Taoist Master* (p. 413). Rochester, Vermont: Inner Traditions.

Bensaude-Bincent, B. (2004). Two cultures of nanotechnology. *HYLE-International Journal for Philosophy of Chemistry, 10*(2), 65–82.

Crichton, M. (2002). *Prey*. New York: Harper Collins.

Drexler, K. E. (1986). *Engines of creation: The coming era of nanotechnology* (p. 14). New York: Anchor Press/Doubleday.

Dupuy, J.-P. (2007). Some pitfalls in the philosophical foundations of nanoethics. *Journal of Medicine and Philosophy, 32*, 237–261.

Dupuy, J. P., & Grinbaum, A. (2004). Living with uncertainty: Toward the ongoing normative assessment of nanotechnology. *Techné, 8*, 4–25.

ETC Group. (2013). Nanotechnology: ETC'S View [DB/OL]. Retrieved July 9, 2013, from http://www.etcgroup.org/issue/nanotechology.

European Commission. (2013). A code of conduct for responsible nanosciences and nanotechnologies research [DB/OL]. A Commission Recommendation of 07/02/2008. Retrieved July 10, 2012, from http://ec.europa.eu/research/consultations/nano-code/consultation_en.htm.

Ferrari, A. (2010). Developments in the debate on nanoethics: Traditional approaches and the need for new kinds of analysis. *Nanoethics, 4*, 27–52.

Gaskell, G., Ten Eyck, T., Jackson, J., & Veltri, G. (2005). Imagining nanotechnology: Cultural support for technological innovation in Europe and the United States. *Public Understanding of Science, 14*, 81–90.

Gordijn, B. (2005). Nanoethics: From utopian dreams and apocalyptic nightmares towards a more balanced view. *Science and Engineering Ethics, 11*(4), 521–533.

Grunwald, A. (1996). Die Bewältigung von Technikkonflikten: Theoretische Möglichkeit und praktische Relevanz einer Ethik der Technik in der Moderne. *Zeitschrift für philosophische Forschung, 51*(3), 437–452.

Jonas, H. (1984). *The imperative of responsibility: In search of an ethics for the technological age* (p. 27). Chicago & London: The University of Chicago Press.

Jotterand, F. (2008). Beyond therapy and enhancement: The alteration of human nature. *Nanoethics, 2*, 15–23.

Joy, B. (2000). Why the future doesn't need us. *Wired, Issue 8*, Apr. Retrieved July 9, 2013, from http://www.wired.com/wired/archive/8.04/joy.html.

Kahan, D. M. (2009). Nanotechnology and society: The evolution of risk perceptions. *Nature Nanotechnology, 4,* 705–706.

Lin, P., & Allhoff, F. (2008). Untangling the debate: The ethics of human enhancement. *Nanoethics*, (2), 251–264.

McGinn, E. R. (2010). What's different, ethically, about nanotechnology: Foundational questions and answers. *Nanoethics, 2,* 115–128.

Moor, J., & Weckert, J. (2004). Nanoethics: Assessing the nanoscale from an ethical point of view. In D. Baird, A. Nordmann, D., J. Schummer, (Eds.), *Discovering the nanoscale* (pp. 305–306). Amsterdam: IOS Press.

Nordmann, A. (2007). If and then: A critique of speculative nanoethic. *Nanoethics, 1*(1), 31–46.

Nordmann, A., & Rip, A. (2009). Mind the gab revisited. *Nature Nanotechnology, 4*(5), 273–274.

Rosenberg, N. (1995). Why technology forecasts often fail. *The Futurist, 29*(4), 16–21.

Schomberg, R. (2007). From the ethics of technology to the ethics of knowledge assessment. In P. Guojon et al., (Eds.), *The information society: Innovation, legitimacy, ethics and democracy* (pp. 39–55). Boston: Springer.

Schummer, J. (2006). Cultural diversity in nanotechnology ethics. *Interdisciplinary Science Reviews, 31*(3), 217–230.

Schummer, J. (2009). Nanotechnologie: Spiele mit Grenzen. Frankfurt a. M.: Suhrkamp.

UNSCO. (2006). *The ethics and politics of nanotechnology* (pp. 17–19). Geneva.

Postscript

The completion of volume relies on many scholars' efforts. More than ten writers contribute their manuscripts, some scholars make an effort to translate, modify and proofread the manuscripts. Several articles included in this volume were initially published in Chinese journals and, with some modifications, translated to English for the first time. What follows is to specify the contribution of this volume.

The introduction was written by myself in Chinese and translated by Li Yang (杨立) and further proofread by me and Mr. Vincent Searfoss, an American scholar teaching in China. Mr. Vincent Searfoss also proofread five articles in "classical readings". Based on the reviewer's opinions about the first manuscript of introduction, much modification and supplement work have been done on this part by me.

The second article "A Brief Discussion of Research in Philosophy of Technology" had been published in *Journal of Northeast Institute of Technology* (Social Science Edition, 1983, Issue 3). It was translated by Xin Liu (刘欣), and further modified by me.

The third article "The Features and Values of Changshu Chen's Philosophy of Technology Ideas" had published in *Journal of Northeastern University* (Social Science Edition, 2016, Issue 3). The translation was finished by Zhaomei Xing (邢兆梅), and was followed by my modification. The reference translation of this article was finished by Yuanyuan Zhang (张媛媛).

The fourth article "Technology and Its Value from the Perspective of Process Theory" was based on the works of Professor Deyu Yuan, and edited and translated by Dongming Cao (曹东溟). Professor Huili Wang (王慧莉) is responsible for modifying this article. It is supplemented by me in the final compilation. The reference translation of this article was finished by Yuanyuan Zhang.

The fifth article "Ethics of Technology and Engineers' Professional Ethics" had been published in Chinese in *Philosophical Researches* (2007, Issue 11). Upon being adopted in this volume, some supplementation work was done. This article was first translated by Jing Zhang (张晶) with my further modification.

The sixth article "The Triad of Science, Technology and Engineering" was edited on the basis of works of Professor Bocong Li. The translation work was

Q. Wang (ed.), *Chinese Philosophy of Technology*, Philosophy of Engineering and Technology 34, https://doi.org/10.1007/978-981-15-1952-9

finished by Li Yang, Xuejiao Li (李雪娇) and Dan Guo (郭聃), and I did further modification.

The seventh article "Aesthetic Factors in Engineering Activity" is the new manuscript of Professor Baowei Zhu. Wei Song (宋薇), Li Yang and Shuang Wang (王爽) took the responsibility for translating this article, which were followed by the modification of Professor Huili Wang. Mr. Leon Whiteley, another American scholar teaching in China, was responsible for proofreading.

The eighth article "The Sinicization of the Philosophy of Technology and the Research on Socialization of Technology" was reorganized based on the previous works of Professor Fan Chen and Professor Jia Chen. Jia Chen translated this article, which were followed by the modification of Professor Huili Wang. Mr. Leon Whiteley was responsible for proofreading.

The ninth article "Reflection on the *Dao of Technology*: Philosophy of Technology from the Perspective of Chinese Culture" written in English by me and Wei Zhang. It was proofread by Xin Liu.

The tenth article "The Developing Process of Technological Rationality and Its Humanistic Relation" was reorganized by Professor Jianjun Zhao on the basis of his previous works. It was translated by Yisu Zhang (张一粟), and proofread by Xin Liu.

The eleventh article "Three Dimensions of Nano-ethics" comes from Professor Sanhu Li on the basis of his previous works. It is translated by Shuang Wang. And it is further modified by Professor Wenyu Liu (刘文宇).

The twelfth article "Technology, Innovation and Life World: A Phenomenological Analysis" was written in English by Professor Dazhou Wang. And the modification was finished by Professor Wenyu Liu.

The thirteenth article "Philosophy of InformationTechnology and Its Contemporary Value" was reorganized by Professor Feng Xiao based on his works before. It was followed by Professor Huili Wang' s modification.

The fourteenth article "Approaches of Nano-ethics and the Chinese Perspective of Ethics of Technology" was finished by Professor Guoyu Wang and Lei Li based on their previous works.

Part of this article was previously published in Chinese: Guoyu Wang, Lei Li, "Four Approaches to Nano-ethical Research", *Journal of Southeast University* (Philosophy and Social Science Edition), vol. 16, No. 1, 2014; Guoyu Wang, "Nanotechnology: From Possibility to Feasibility", *Philosophical Researches*, No. 8, 2011. Xiaoling Sui (隋晓玲) was in charge of the translation work, which was followed proofread by Professor Wenyu Liu.

The professors and associate professors engaged in modifying this volume come from School of Foreign Languages, Dalian University of Technology and Dalian University of Foreign Languages. Their efforts are highly appreciated.

Qian Wang

Index

Q. Wang (ed.), *Chinese Philosophy of Technology*, Philosophy of Engineering and Technology 34, https://doi.org/10.1007/978-981-15-1952-9

The manufacturer's authorised representative in the EU is Springer Nature Customer Service Centre GmbH, Europaplatz 3, 69115 Heidelberg, Germany. If you have any concerns regarding our products, please contact ProductSafety@springernature.com

Printed and bound by CPI Group (UK) Ltd, Croydon, CR0 4YY
25/09/2026
02222898-0005